Beginner's (

Digital Electr\

Beginner's Guides are available on the following subjects:

Amateur Radio
Audio
Basic Programming
Building Construction
Cameras
Central Heating
Colour Television
Computers
Domestic Plumbing
Electric Wiring
Electronics
Fabric Dyeing and Printing
Gemmology
Home Energy Saving
Integrated Circuits
Microprocessors
Photography
Processing and Printing
Radio
Spinning
Super 8 Film Making
Tape Recording
Technical Illustration
Technical Writing
Television
Transistors
Video
Weaving
Woodturning
Woodworking

Beginner's Guide to

Digital Electronics

Ian R. Sinclair

Newnes Technical Books

Newnes Technical Books
is an imprint of the Butterworth Group
which has principal offices in
London, Boston, Durban, Singapore, Sydney, Toronto, Wellington

First published 1980
Reprinted 1983

British Library Cataloguing in Publication Data

Sinclair, Ian Robertson
  Beginner's guide to digital electronics.
  1. Digital electronics
  I. Title
  621.3815   TK7868.D5   79-41414

  ISBN 0-408-00449-5

Photoset by Butterworths Litho Preparation Department
Printed and bound in Great Britain by The Thetford Press Ltd,
Thetford, Norfolk

Preface

Even a beginner's guide must assume that the reader has some previous knowledge, and this book, dealing as it does with some very up-to-date technology, can be no exception. However, only some knowledge of transistor and f.e.t. operation, and of linear circuit components, has been assumed; otherwise the book is a step-by-step introduction to digital techniques, with practical hints.

Nowadays no book claiming to deal with digital electronics can omit mention of the microprocessor, and the final chapter of this book is concerned entirely with this component.

Mathematical topics have been reduced to a minimum but a guide to Boolean algebra is included for the use of those who may be beginners as far as digital circuits are concerned but whose knowledge of mathematics has reached a more advanced stage. The digital symbols used in the circuits are the internationally recognised MIL-806B standards, rather than the recently changed and unfamiliar British Standards BS3939: 1977. Although the BS symbols are used for City and Guilds and TEC examinations in electronics they are not likely to be found anywhere else; accordingly, they have been omitted from this work as their appearance would be confusing.

I.R.S.

Contents

1 Signals, switching and devices

An electronic signal consists of a changing voltage which can be used to represent some non-electronic quantity such as a sound wave, a picture, a number, a speed etc. The *waveshape*

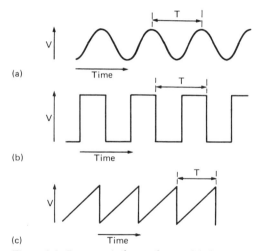

Figure 1.1. Some typical waveshapes: (a) sinewave, (b) rectangular wave, or square wave, (c) sawtooth. The time period T is measured between two consecutive identical peaks of the wave

is the name given to the appearance of a graph of voltage plotted against time for such a varying voltage. Some typical wave-shapes are shown in *Figure 1.1*. When a wave contains a repeating pattern, as all the waveforms of *Figure 1.1* do, the time for one complete wave is called the *time period* (*T*) and the inverse of this quantity (1/*T*) is called the *frequency (f)*.

The *amplitude* of the wave is its size in volts, and the most useful measurement is *peak-to-peak* amplitude, as indicated in *Figure 1.2*. Measurement of peak amplitude and r.m.s. amplitude is of little interest in digital electronics.

Figure 1.2. Peak-to-peak amplitude. This is the most useful type of measurement for digital signals and it is made with an oscilloscope

Sound waves and televised pictures can be turned into waveforms in which the shape of the wave is very important, as important as the frequency or the amplitude. Signals of this type are called *analogue* signals: the word analogue implies that there is a strict correspondence between the size (amplitude) of, for example, the sound wave and the electrical wave generated from it. Any operation upon an analogue signal which changes the shape of the waveform is therefore causing distortion, losing some of the signal information.

Digital signals convey information using entirely different principles. As the name suggests, digital signals represent

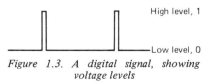

Figure 1.3. A digital signal, showing voltage levels

numbers and are a coded way of carrying information, just as Morse code is a coded method of communicating words. The waveform of a digital signal is usually that of a steep-sided pulse (*Figure 1.3*), but the precise shape is unimportant. The only important features of a digital waveform are the two levels between which the voltage changes. These can be called 'high' and 'low', 'on' and 'off', but are more usually labelled as 1 and 0.

Since the waveshape of a digital signal is unimportant, linear amplifiers are not needed for digital signals, and the techniques of negative feedback which are so much used for linear amplifiers are irrelevant. The precise amplitude of signal is also unimportant, provided that the two voltage levels (0 and 1) are quite different — we can arrange the circuits so that a voltage near to the 0 level will be taken as 0, and a voltage near the 1 level will be taken as 1.

Two portions of the digital signal need to be of a specified shape however, and these are the leading and trailing edges. An

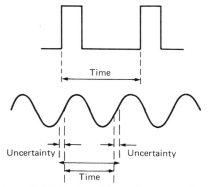

Figure 1.4. Why fast-rising pulses are used — a time interval can be measured precisely between sharp pulses, but there is always a considerable uncertainty when slow-changing waves are used

ideal digital wave would be a rectangular pulse, with the voltage rising from 0 to 1 instantly and falling from 1 to 0 equally swiftly. In practice, instantaneous changes are impossible, but rise and fall times measured in tens of nanoseconds (1 nanosecond = 10^{-9}s, one thousandth of a millionth of a second) are practical values. These rapid rises and falls (the rise time is usually more important) are desirable for two reasons, the first being that digital signals are often used for precise time measurement, and a rapidly rising voltage is a good starting point; a slowly rising voltage would introduce considerable uncertainty (*Figure 1.4*) into a time reading. The second reason for using fast rising and falling pulses is that many digital circuits contain high-gain amplifiers which are either switched off or fully on,

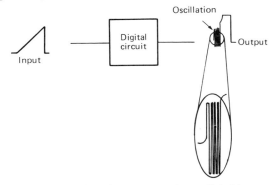

Figure 1.5. A slow-changing wave into a digital i.c. can cause oscillation during the output signal

and which can oscillate if they are left for any significant time biased between the 'off' and 'on' settings, as they would be in the middle of a slowly changing wave (*Figure 1.5*). A significant time as far as starting an oscillation is concerned would be a microsecond or so.

Logic and counting

Digital circuits are used for many applications, but the types of application classed as *logic* or *counting* are by far the most important. Logic circuits use the two levels 1 and 0 to mean 'yes' or 'no' respectively as the answer to a question which can have only two possible answers. Any process which can be broken down to a series of tests, all of which have yes/no answers, can be carried out by digital logic. For example, consider the coin-selecting mechanism of a vending machine. The first test may be of coin diameter — 'too large or not'? If the answer is 0 for 'not' then the second test must be 'correct or too small'? By breaking this measurement down into two go/nogo tests in this way, digital logic can be applied, and similar tests for weight, metal type, edge milling etc. can be carried out.

Any logic process can be broken down into a series of simple steps, each of which will have a 'yes' or 'no' outcome which can be represented by a 1 or 0 output signal. We shall see in the next chapter how such systems can be designed and analysed.

Even more important than logic circuits is the application of digital methods of counting. The use of two voltage levels means that only the figures 0 and 1 can be represented, so a counting scale using only these figures must be used, and the *binary scale* is one such counting scale.

The binary scale represents numbers, using only the figures 0 and 1, in the same way as the everyday scale-of-10 (denary or decimal scale) uses the figures 0 to 9. When we write a number such as 684 we interpret it as six hundreds plus eight tens plus four units, and the separation of the columns for units, tens and hundreds makes the operations of addition, subtraction, multiplication and division extremely simple. To be fully persuaded of this, try adding the Roman numerals CLXII + CCXXIV *without* converting to denary scale.

In the *denary scale* each column represents a power of 10 by which the figure in the column is multiplied. The power is the number which shows how many times 10 has to be multiplied by itself to obtain the column value. For example, when the power of 10 is 2 then the quantity written as 10^2 is 100, and the column is the hundreds column. Similarly, the third power of 10 gives 10^3 or 1000, and by convention we take $10^1 = 10$ and $10^0 = 1$. We can now number our figure columns with powers of 10 as in *Figure 1.6* using 0 for the units column, 1 for the tens column, 2 for the hundreds column and so on. The figure in the 10^0 column (units) is known as the *least significant digit*; the one in the highest power column is known as the *most significant digit*. The reason for these names is fairly obvious; an error of one or more digits is much more serious in the highest power column than in the lowest power column.

Denary number

10^3	10^2	10^1	10^0
1000	100	10	units

| 1 | 7 | 4 | 2 |

one   thousand
seven   hundreds
four   tens
two   units

Binary number

2^4	2^3	2^2	2^1	2^0
16	8	4	2	1

| 1 | 0 | 1 | 1 | 0 |

1   sixteen
0   eight
1   four
1   two
0   one

Figure 1.6. Denary (scale-of-10) and binary (scale-of-2) numbers. Each place to the left in a number represents a higher power of the base number. The base of denary numbers is 10, the base of binary numbers is 2

A scale of 2 can be drawn up in exactly the same way using columns representing 2^0, 2^1, 2^2, 2^3, 2^4 and so on. The figure placed in each column must be either a 0 or a 1, since these are the only figures we have. Because any quantity multiplied by 0 is 0, and any quantity multiplied by 1 is that quantity unchanged, the conversion of a binary number into denary is simple: see *Figure 1.7 (a)*.

For example, 101101 has a 1 in each of the columns 2^0, 2^2, 2^3 and 2^5 and the value of 101101 in denary must therefore be $2^5 + 2^3 + 2^2 + 2^0$. Now, $2^5 = 32$, $2^3 = 8$, $2^2 = 4$, $2^0 = 1$ so that 101101 is equivalent to $32 + 8 + 4 + 1 = 45$ in denary. *Table 1.1* shows the values of powers of two up to 2^{23}. Converting denary whole numbers to binary is most easily done using the method shown in *Figure 1.7 (b)*. *Figure 1.8* illustrates how the methods shown above may be extended to fractions, using negative powers to represent quantities such as 0.1 (10^{-1}), 0.01 (10^{-2}) and their binary counterparts 0.5 (2^{-1}), 0.25 (2^{-2}) and so on.

When a number is written in the binary scale, each digit is called a *bit* (short for *b*inary dig*it*). Groups of eight such bits

Table 1.1   POWERS OF TWO

2^0	1	2^8	256	2^{16}	65 536
2^1	2	2^9	512	2^{17}	131 072
2^2	4	2^{10}	1 024	2^{18}	262 144
2^3	8	2^{11}	2 048	2^{19}	524 288
2^4	16	2^{12}	4 096	2^{20}	1 048 576
2^5	32	2^{13}	8 192	2^{21}	2 097 152
2^6	64	2^{14}	16 384	2^{22}	4 194 304
2^7	128	2^{15}	32 768	2^{23}	8 388 608

Note: In computing parlance, 1024 ($= 2^{10}$) is known as 1K, as distinct from the usual use of 1k as 1000 in electronics work. So 8K is 8192 (not 8000), for example.

are very commonly used in counting, particularly in micro-processors, and are called a *byte*.

The bit which represents 2^0 (0 or 1) is called the *least significant bit* (LSB), and the bit which represents 2^7 (128) is called the *most significant bit* (MSB) of a single byte.

The simple binary code is not the only possible way of using the digits 1 and 0 to represent numbers. The *Gray code* is used in devices which convert analogue quantities to digital signals, because it is more error-free. In a Gray code count, only one digit changes when the count is increased by 1 (*Table 1.2*), but because arithmetic in Gray code is difficult, Gray code numbers are always converted to binary for processing.

Binary Coded Decimal systems (BCD) use four binary digits to represent each digit of a decimal number. This can be done by the 8-4-2-1 code, usually referred to as BCD, in which each decimal digit is converted to its four-bit binary equivalent, as in *Figure 1.9*, but another system which is sometimes used is the *Excess-3* code, in which 3 is added to each digit before converting, as shown in *Figure 1.10*. The advantage of the Excess-3 code as compared to the 8-4-2-1 system is that addition and subtraction can be carried out more easily.

(1) Write down binary number, and note place number (equal to power of 2) for each 1 in the number. Place numbers start with 0 at the right-hand side:

$$\begin{array}{cccccc} & 9 & 6 & & 3\,2 & 0 \\ \text{Example:} & 1\ 0\ 0\ 1\ 0\ 0\ 1\ 1\ 0\ 1 \end{array}$$

(2) Now write value of each power of 2 you have noted, and arrange in a column:

$$\begin{array}{rl} \text{Example:} & 2^0 \quad 1 \\ & 2^2 \quad 4 \\ & 2^3 \quad 8 \\ & 2^6 \quad 64 \\ & 2^9 \ 512 \end{array}$$

(3) ... and add: $\overline{589}$

Denary equivalent of 1001001101 is 589

Figure 1.7 (a). Converting from binary numbers to denary

Number to be converted: 1742

Remainders

$$\begin{array}{ll} 2)1742 & \\ 2)871 & 0 \\ 2)435 & 1 \\ 2)217 & 1 \\ 2)108 & 1 \\ 2)54 & 0 \\ 2)27 & 0 \\ 2)13 & 1 \\ 2)6 & 1 \\ 2)3 & 0 \\ 2)1 & 1 \\ 0 & 1 \longleftarrow \text{Read binary number from here} \end{array}$$

Binary equivalent of 1742   1 1 0 1 1 0 0 1 1 1 0

Figure 1.7 (b). Converting from denary numbers to binary. The denary number is divided by 2 and the remainder noted. This action is repeated until the final division which always leaves a remainder of 1. The remainders are then read in opposite order, last first, to form the binary number

Fractional powers

| Denary fraction .1111: | | | | Binary fraction .1011: | | | |

10^{-1}	10^{-2}	10^{-3}	10^{-4}	.5	.25	.125	.0625
1	1	1	1	2^{-1}	2^{-2}	2^{-3}	2^{-4}
				. 1	0	1	1

= one tenth
+ one hundredth
+ one thousandth
+ one ten thousandth

= 1 × .5
+ 0 × .25
+ 1 × .125
+ 1 × .0625

= 0.6875 (in denary)

Converting denary fractions to binary

Rules: Multiply by 2. Count a 1 on the left-hand side of the decimal point as a binary 1, count a 0 or 2 as a binary zero. Continue for as long as needed — many binary fractions never work out finally. Read binary number from the top downwards.

Example			0.624	Denary
			×2	
	1	◄———	1.248	
			.248	
			×2	
	0	◄———	0.496	
			×2	
	0	◄———	0.992	
			×2	
	1	◄———	1.984	
			.984	
			×2	
	1	◄———	1.968	
			.968	
			×2	
	1	◄———	1.936	

Binary fraction is .100111 to six places (actually equivalent to 0.596 in denary)

Figure 1.8. Fractions — a binary fraction uses negative powers of 2 after the point, just as a denary fraction uses negative powers of 10. The conversion of denary fractions into binary fractions is seldom exact

Table 1.2 GRAY CODE

Decimal	Gray code	Decimal	Gray code
0	0000	8	1100
1	0001	9	1101
2	0011	10	1111
3	0010	11	1110
4	0110	12	1010
5	0111	13	1011
6	0101	14	1001
7	0100		

Note: The value of the Gray code is that only one digit changes at each unit of a count. This avoids errors when some types of analogue-to-digital conversions are carried out − for example, shaft position encoders. The Gray code has to be converted to binary for arithmetic operations, and i.c.s exist which can carry out the conversion.

Denary number: 167

In binary, 1 is 0001
      6 is 0110   } using 4 bits for each figure
      7 is 0111

167 is   <u>000101100111</u> in BCD

Note: In *binary*, 167 is   <u>10100111</u>

Figure 1.9. The BCD 8-4-2-1 code uses four bits to represent each figure of a decimal number. This is particularly suitable for displaying denary numbers (see Chapter 6)

Denary number: 257

Add 3 to each:	5	8	10
Then into 4-bit binary:	0101	1000	1010
Excess-3 number is	<u>010110001010</u>		

Figure 1.10 The Excess-3 code

Simple binary arithmetic

The operations of binary arithmetic can be carried out, on paper, in exactly the same way as those of denary arithmetic. Addition and subtraction are illustrated in *Figure 1.11*, using

Addition	Rules
1 1 1 1 1 1 ← carry	$0 + 0 = 0$
1 0 0 1 1 0 1	$1 + 0 = 1$
0 1 1 1 0 1 1	$1 + 1 = 0$ carry 1
= $\overline{10001000}$	carry $1 + 1 + 1 = 1$ carry 1

Subtraction	Rules
1 1 0 0 1 1 0 1 0	$0 - 0 = 0$
− 0 1 1 0 1 0 1 0 1	$1 - 0 = 0$
= $\overline{011000101}$	$1 - 1 = 0$
	$0 - 1 = 1$ carry 1
	$1 - 1 - $ carry $1 = 1$ carry 1

Figure 1.11. Addition and subtraction in binary numbers

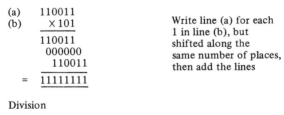

Multiplication

(a)   110011
(b)   × 101
             ──────
             110011
             000000
             110011
             ────────
   =   11111111

Write line (a) for each
1 in line (b), but
shifted along the
same number of places,
then add the lines

Division

```
        1001
110 )110110
     110
     . . .
        110
        110
```

Use exactly the same
method as is used for
denary long division

Figure 1.12. Multiplication and division in binary numbers

12

Number:	1 0 0 1 0 1 1 1
1's complement	0 1 1 0 1 0 0 0
2's complement	+ 1
	0 1 1 0 1 0 0 1

Subtraction using 2's
complement:

  1 1 0 0 1 0 0 (a)
− 0 1 1 0 0 1 1 (b)

Note:

Take 2's complement of (b)

− 1 0 0 1 1 0 1

and add to (a)

  1 1 0 0 1 0 0
  1 0 0 1 1 0 1
  <u>10110001</u>

and *discard* the final carry
to get

<u>0 1 1 0 0 0 1</u>

For   1 1 0 0 0 1 1 0 1
    −   1 0 1 1 1

rewrite as 1 1 0 0 0 1 1 0 1
    − 0 0 0 0 1 0 1 1 1

which is 1 1 0 0 0 1 1 0 1
    + 1 1 1 1 0 1 0 0 1
    (1) <u>1 0 1 1 1 0 1 1 0</u>

Answer: <u>1 0 1 1 1 0 1 1 0</u>

Figure 1.13. Two's complement arithmetic. The number to be subtracted is 2's complemented and the complement then added to the other number. Both numbers must contain the same number of digits (adding 0s before the first figure of the number if necessary)

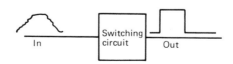

Figure 1.14. Using a switching circuit to 'sharpen-up' a pulse

the carry when a sum exceeds 1 or when subtracting 1 from 0. Multiplication and division are carried out as shown in *Figure 1.12*. These methods are not all particularly easy to apply to digital circuits. Circuits which perform addition will be dealt with later, and a simple modification of arithmetic makes subtraction possible without using a different circuit. The method used is called *2's complement subtraction* and consists of the following process:

(1) The 1's complement of the number which is to be subtracted is formed by exchanging 1s for 0s and 0s for 1s (*Figure 1.13*).
(2) Another 1 is added to the least significant bit (LSB).
(3) The 2's complement is now *added* to the other number. By using this method, the same circuit which performs addition will also perform subtraction, provided that the 2's complement of the number being subtracted can be formed. As we shall see later, forming a 2's complement is not a difficult circuit task. Multiplication and division are carried out using shift registers (Chapter 7) in addition to adding circuits.

Digital switching circuits

A switching circuit is one whose output voltage shifts rapidly from one extreme to another when a digital signal is applied to the input. Digital circuits make use of switching in the way that linear circuits make use of amplification, with the important difference that the input and output signal amplitudes of a switching circuit are very often identical. Another important difference is that the switching circuit will act to 'sharpen up' the leading and trailing edges of a pulse which has become smoothed out (integrated) in a cable (*Figure 1.14*).

A simple voltage switching circuit is shown in *Figure 1.15*. The transistor is a bipolar type (in this case, *n-p-n*), and uses a 330 ohm load resistor in its collector circuit. The base circuit is unbiased, and connected to the input through a 4.7 kΩ resistor

R1 which limits the amount of current that can flow in the base circuit. A 5.0 V supply line voltage is assumed since this is a standard voltage for many digital circuits.

For an input voltage of less than about 0.5 V, the transistor remains cut-off, so that the collector voltage, which is the output voltage of the circuit, remains high. Any voltage between 0 V and +0.5 V will therefore count as a 0 input. When the base voltage exceeds +0.5 V, the transistor will switch on; but the digital signal applied to the input will not remain at 0.5 V but change abruptly from about 0 V to around +5.0 V. With +5.0 V at the input, the 4.7 kΩ input resistor, along with the base input resistance of the transistor (perhaps a few hundred ohms) will allow about 1 mA to flow into the base (using Ohm's law and assuming a total of 5.0 kΩ resistance).

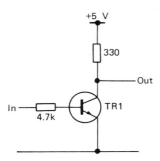

Figure 1.15. A simple switching circuit

Now the maximum amount of current that can flow in the collector circuit is limited by the supply voltage and the collector resistor. If we assume that the collector voltage of the transistor could drop to 0 V, then the maximum collector current would be 5.0/0.33 = 15.15 mA. To produce this amount of current in the collector circuit with 1 mA flowing in the base circuit needs a transistor with a current gain of at least 15, a value which is easily exceeded. A normal 1 input will therefore ensure that the transistor saturates — passing as much current as it can in the collector circuit, so that the collector voltage drops to a very low value, around 0.2 V. The excess current flowing in the base circuit ensures that the transistor saturates even if the 1 voltage is less than the 5.0 V we have assumed.

Switching can also be carried out by a MOSFET, as indicated in *Figure 1.16*. No limiting resistor is needed, because no current flows in the gate circuit, but a resistor or diode must be connected between gate and earth to prevent the gate from being affected by 'stray' electrostatic voltages. Once again, the amount of current which flows in the drain-to-source circuit is limited by the load resistor, which is usually of several kΩ. Once again, a small change of input voltage causes the output to change between 1 and 0. Ideally, the circuit is arranged so that any voltage below +1.0 V counts as a 0 and any voltage above 10.5 V (assuming 12.0 V supply) counts as 1.

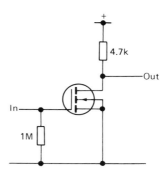

Figure 1.16. A simple MOS switching circuit

The circuits of *Figure 1.15* and *1.16* have, for the sake of simplicity, shown a single stage switching circuit. Digital circuits generally use switching circuits consisting of several stages, however, and have rather more power gain than a single stage. The reasons for using several stages are:

(1) A high gain will ensure that the switchover is fast.
(2) A high gain will ensure that the switchover is complete for only a small voltage change at the input.
(3) The output stage of the switching circuit can be arranged so that it can provide enough signal power to drive several other switching circuits.

Treating these reasons in more detail, *Figure 1.17* shows a typical input/output graph for a switching circuit which uses a 5 V supply.

Imagine an input signal which is a voltage rising from 0 V to +5 V in a time of 1 μs, as indicated. The rise time of this voltage is defined as the time needed for the voltage to change from 10% to 90% of its final value (5.0 V in this example). This time, from 0.5 V to 4.5 V, is the time to rise by 4.0 V and is $^4/_5$ μs, 0.8 μs. Now the input/output graph shows that the output

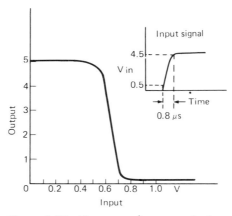

Figure 1.17. The output/input graph for a switching circuit, and how it causes a signal to be sharpened

changes from 4.5 V to 0.5 V for an input change of 0.6 V to 0.7 V. How long does this take? If the input changes by 5 V in 1 μs, then it changes by 0.1 V (0.7−0.6 V) in $\dfrac{1.0 \times 0.1}{5.0}$ μs, which is 0.02 μs. We would expect that the output change of 4.5 V to 0.5 V would also take this time, so that the fall time at the output is very much faster than the rise time at the input.

This, however, is true only if the output of the switching circuit can charge or discharge the stray capacitances in the circuit at this speed. In addition, this switching circuit may have to switch on or off several other switching circuits, each of which will have its own stray capacitance and will also require some input current.

At one time, switching circuits for digital systems used separately wired transistors, an arrangement called a *discrete component circuit*. However, as digital circuits proved their usefulness and demand grew the number of transistors, and even worse the number of soldered joints, in a digital circuit became excessive. The use of integrated circuits is an answer to both of these problems. An *integrated circuit* (i.c.) is a complete circuit of transistors and resistors which is constructed on a tiny chip of silicon, using the same methods as are used to construct transistors. The first i.c.s contained only a few transistors and resistors, and are known as SSI (Small Scale Integration) circuits, but as the technology developed, it soon became possible to construct i.c.s containing the equivalent of 50 or more transistors. These i.c.s became known as MSI (Medium Scale Integration) circuits. Later it was discovered that MOSFET circuits could be fabricated as i.c.s and very large numbers of f.e.t.s could be packed on one chip. These LSI circuits (Large Scale Integration) can contain thousands of transistors and resistors arranged into circuits which would be much too costly and too complex to reproduce using conventional construction methods.

The availability of i.c.s has led to much greater use of digital circuits for applications which previously were carried out by either linear circuits or non-electronic methods. Because the i.c. is made in one sequence of operations similar to those used for a transistor, it is as easy and cheap to make an i.c., even an LSI chip, as to make a transistor. The design work on the i.c. is incomparably more expensive but this can be recovered if enough i.c.s are made and sold. Reliability is improved because the i.c. can be tested and its use in place of the separate components it displaces means that fewer connections have to be made. Most noticeable of all, the size of a circuit can be greatly reduced, so that it is now possible to design digital equipment which will fit inside other components. For example, the circuitry needed to convert a TV receiver into a miniature computer can easily be added and placed *inside* the receiver's cabinet. Before LSI, the circuitry for the computer would have occupied most of the living room. For more details of i.c.s see the companion *Beginner's Guide to Integrated Circuits*.

With a few exceptions, i.c.s used for digital circuits are of two types: TTL and MOS. The TTL types use conventional (bipolar) transistors in integrated form; the MOS, as the name suggests, use MOSFETs in integrated form, usually with both *p*-channel and *n*-channel types on one chip. Though both 'families' contain similar circuit functions, the differences between the two types are important and we shall look at each type in more detail.

TTL digital i.c.s

The name 'TTL' is an acronym for Transistor-Transistor-Logic, a scheme which has replaced the earlier DTL (Diode-Transistor-Logic) and RTL (Resistor-Transistor-Logic) circuits. TTL circuits, of which the 74 series manufactured by Texas Instruments is the best-known example, are MSI circuits which make

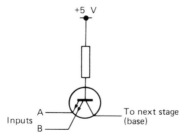

Figure 1.18. The input of a typical TTL circuit is to one emitter of a transistor whose base is returned through a resistor to the positive supply voltage

use of *n-p-n* transistor structures. The operating voltage is 5.0 V and the design of the circuits is such that this voltage must not be exceeded.

A typical TTL input stage is shown in *Figure 1.18*. The input transistor has its base connected to a resistor which in

turn is connected to the +5.0 V supply. The input to the TTL
circuit is to the *emitter* of this transistor, not to the base, so
that this first stage is a common-base switching circuit. When
the input voltage at the emitter is high, between 4.5 and 5.0 V,
the first transistor will not conduct because the voltage between
base and emitter is not high enough. No current will flow
between collector and emitter.

When the voltage at the emitter of the first transistor is low,
near 0 volts, current will flow between the base and the emitter.
In the standard series of TTL circuits, this current is set at
1.6 mA by the value of the resistor which connects the base to
the +5.0 V supply. This current is enough to saturate the first
transistor, meaning that the collector-to-emitter path is low
resistance, as low as can possibly be obtained. Because the
emitter of this transistor is set at 0 V, then, the collector
voltage will also be low, no more than 0.2 V above the emitter
voltage.

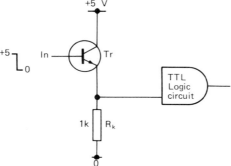

*Figure 1.19. Driving a TTL circuit from an
emitter-follower — unless R_k is small the TTL
circuit will not be switched off at the input.
Values of R_k of 470Ω or less are needed*

Because of this construction, examples of which are illustrated in Chapter 2, the input of any TTL circuit must be
driven from a low impedance source, capable of passing 1.6 mA
at a low voltage. Imagine, for example, a TTL circuit driven
from an emitter follower (*Figure 1.19*). With the emitter

follower biased on, the TTL input would be at logic 1, but switching off the emitter follower would not necessarily turn off the TTL circuit. The reason is that the resistor R_k might be of too high a value to let 1.6 mA flow. For example, with $R_k = 1$ kΩ, 1.6 mA would cause a drop of 1.6 V, which might not be low enough to let the TTL circuit switch over. TTL inputs must therefore be driven from circuits which will allow currents of at least 1.6 mA to flow to earth with a very small voltage drop. A suitable circuit is, for example, the common emitter circuit of *Figure 1.20*. When the base voltage of the

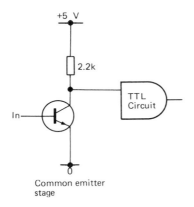

Common emitter stage

Figure 1.20. Driving a TTL circuit from a common-emitter amplifier – a much more satisfactory way of using a transistor to drive TTL circuits

transistor is raised to around 0.6 V, the transistor saturates, so that the resistance between collector and emitter is very low, and the TTL input is held to about +0.2 V. This is certainly low enough to ensure that the TTL input is switched over to 0. Note, incidentally, that if a TTL input is not connected, it will 'float' to a permanent high voltage, logic 1, signal. This should not, however, be relied on as a method of keeping an input at logic 1 unless the digital signals are of low frequency (100 Hz or less). At high frequencies the capacitive coupling between one input and others may be enough to cause unwanted input signals to a disconnected pin, so unused inputs should be connected through a 1 kΩ resistor to +5 V, or directly connected to earth if a 0 input is needed.

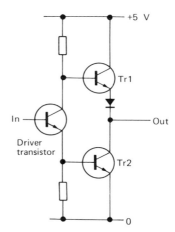

Figure 1.21. A typical TTL output circuit. A few TTL circuits omit Tr1 to form what is called an 'open-collector' output stage (see Chapter 6)

The type of output stage which is used in TTL circuits is illustrated in *Figure 1.21*. Two *n-p-n* transistors are connected in series and the output terminal is connected at the point where the emitter of one transistor joins the collector of the other. To switch this output to logic 1, transistor Tr1 has its base voltage taken high, to +5.0 V, and transistor Tr2 simultaneously has its base voltage taken low, to 0 V. The output terminal now has a low resistance to the +5.0 V supply line, so that current can pass through a load connected between the output terminal and earth. In this condition, the output terminal can act as a source of current.

When the circuit switches over, Tr1 is cut off, with its base voltage low, and Tr2 is saturated, with its base voltage held at about 0.6 V. In this state the output terminal is at low voltage logic 0, and current can pass into the output terminal from a load connected to the +5.0 V line. In this condition, therefore, the output terminal can act as a sink for current.

In either logic state, a current passing out of or into the output terminal causes very little change in the voltage at the output terminal. For many TTL circuits, the maximum current which can pass into or out of the output terminal is guaranteed as 16 mA. In digital designer's language the output stage can source or sink a maximum of 16 mA. The importance of this is that the low impedance and the current handling capacity of this output stage enables us to use TTL output to drive other circuits. Low current relays, for example, can be driven directly from a TTL output (subject to the usual diode protection against the voltage surge occurring when current through a relay is switched off) and smaller loads such as l.e.d. indicators can all be driven.

More important, the current sinking ability of an output stage enables one TTL output to drive up to 10 TTL inputs, each of which needs 1.6 mA to keep the voltage down to logic 0. In digital designer's language the TTL output stage has a 'fanout' of 10.

TTL circuits are used in mainframe (large) computers because of one considerable advantage — very fast operating speeds. The time needed for a TTL circuit to switch from 0 to 1 or 1 to 0 is about 10 ns, depending on the type of circuit, and even faster speeds can be obtained using the 74H series of TTL circuits. The price which has to be paid for high-speed operation is considerable power dissipation, so that TTL circuits need large-current 5.0 V power supplies. The use of a modified design (Schottky) of TTL, however, enables lower operating currents to be used, and the 74LS series of TTL circuits (the LS signifying low power Schottky) can also achieve high operating speeds because the transistors are not allowed to saturate. The principle which permits this is the construction in integrated form of a Schottky barrier diode between the base and the collector of each transistor. A Schottky diode uses an aluminium-silicon junction and has a very low forward voltage when conducting, of the order of 0.3 V, so that such a diode connected between the base and the collector (*Figure 1.22*) of a transistor prevents the collector voltage from falling to less than 0.3 V below the base voltage. In this way, when a transistor is switched on, excessive current flows through the Schottky diode from base to collector (and so the earth) instead of causing saturation. Low power Schottky TTL i.c.s typically dissipate only one fifth of the power required by the standard

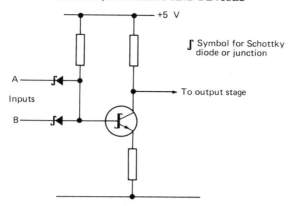

Figure 1.22. A typical Schottky low-power input stage. The logic switching is done by the diodes, and the transistor, which also uses Schottky junctions, performs power amplification. Switching speed can be very high, with much lower currents flowing

TTL circuits; alternatively, the use of Schottky circuits operating at higher power levels permits switching times of around 3 ns.

CMOS circuits

Large scale integration requires circuits which dissipate very small amounts of power, since thousands of transistors have to be accommodated on one chip. The use of MOS construction enables switching circuits with very low *p*- and *n*-channel MOSFETs, known as *Complementary MOS* (CMOS) and pioneered by RCA, has resulted in a complete family of logic circuits.

A typical circuit of a MOS digital i.c. is shown in *Figure 1.23*. At the input two f.e.t.s are connected in series, one *p*-channel and the other *n*-channel. The gates are connected together, so that the same signal is applied to each, and a network of diodes and resistors is arranged to protect the thin gate insulation from excessive voltage at the input. The action of the circuit is that a positive voltage at the input will allow current to flow in the n-channel f.e.t., so making a low resistance connection between the output terminal and earth. Conversely, a 0 voltage at the input will allow current to flow in the p-channel f.e.t., so making a low resistance connection between the output terminal and the positive supply line. The output terminal is

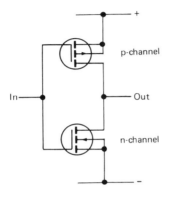

Figure 1.23. A typical CMOS inverter circuit, omitting the gate-protection diodes

connected to both drains of the f.e.t.s, so that the f.e.t.s act as voltage amplifiers in each direction. Several circuits use only one set of f.e.t.s with no separation of input and output stages, so that the internal circuitry is simpler than that of TTL circuits. The operating time is longer than TTL, typically $35-100$ ns, but the power dissipation is negligible.

In addition, the gate input takes no measurable current either at the logic 1 or logic 0 voltages, so that the fanout which can be achieved is limited only by the stray capacitance of the number of stages which are connected. A fanout of 50 is typical for many CMOS circuits. On the other hand, the current sinking or sourcing ability of the output stage is limited to about 0.5 mA, so that interface circuits are needed if any loads apart from other CMOS circuits are to be driven. Even an l.e.d. indicator, for example, may have to be driven by an emitter follower whose base is driven from the CMOS circuit.

CMOS digital circuits can be operated from power supplies ranging from +3.0 V to +12.0 V, so that the 5.0 V of the TTL circuits is compatible. In addition, the low consumption of CMOS circuits permits battery operation, so enabling the popular 6.0 V or 9.0 V transistor radio batteries to be used.

For most purposes, excluding mainframe computers, CMOS circuits are fast enough and their tolerance of different supply voltages makes them useful for a variety of circuits. For experimental use, however, CMOS circuits suffer from the disadvantage of being readily damaged by stray voltages because of the high resistance of the gate terminals. The electrostatic voltage on insulators (including printed circuit boards) or on human hands can be enough to damage CMOS gates, and the protection circuits are not effective until the i.c. has been connected into circuit. The most important precautions are:

(1) To connect the i.c.s into circuit only when all components have been connected. All of the pins must be connected into the circuit.

(2) To use i.c. holders wherever possible.

(3) To use soldering irons which have earthed metal work, and to solder the earth and supply pins first when connecting the i.c.

(4) To keep the i.c. in its protective plastic foam until it is inserted.

(5) To avoid touching the pins of the i.c.

(6) To treat a complete p.c.b. as carefully as the i.c. itself until the earth line is permanently connected.

Once connected, a CMOS i.c. is as electrically robust as any other i.c; the only vulnerable period is that between removing the i.c. from its protective plastic foam (a conducting material) and connecting it into its circuit. For this reason, though, TTL i.c.s are preferable for breadboarding or patchboard work when pins may be open-circuited during changes of wiring.

Practical work

Like other branches of electronics, digital electronics requires some practical work for complete understanding. The digital circuits illustrated in this book can be tested in practice using either TTL or CMOS i.c.s, but the use of TTL will be more convenient in most cases provided that a stabilised 5.0 V supply is available. If only 9.0 V battery power can be obtained, CMOS i.c.s can be used, but subject to the handling precautions mentioned earlier.

One considerable practical advantage of working with digital circuits is that very low-speed operation is possible, so that signal levels can be studied without the need for an oscilloscope. Since the signal levels are either 1 or 0, the main diagnostic instrument is simply an l.e.d. which glows when connected to 1 and not when connected to 0. To avoid excessive current, a resistor must be connected in series when the l.e.d. is used with TTL circuits (*Figure 1.24*) and a combination of transistor and resistor used for CMOS (*Figure 1.25*).

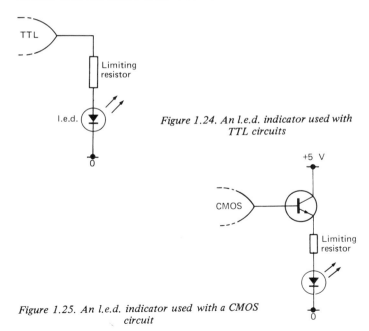

Figure 1.24. An l.e.d. indicator used with TTL circuits

Figure 1.25. An l.e.d. indicator used with a CMOS circuit

To avoid continual soldering and desoldering, a solderless breadboard form of construction is recommended. Several types are available, but the least expensive and the most convenient to use is the Eurobreadboard (*Figure 1.26*). This has four columns of contacts, each column consisting of 25 rows of five contacts each. All the contacts in a row are connected, and the columns are so arranged that any digital i.c. package

Figure 1.26. The Eurobreadboard. This is an ideal way of trying out digital circuits since construction on it is very rapid and all the components are easily replaceable and re-usable

can be placed on the board. The compact shape of the board ensures that long wire leads are not needed when connections are made from one i.c. to another. For ease of construction the rows and columns are numbered and lettered respectively, so that all connections can be made by reference to this indexing system before the i.c.s are inserted. This enables CMOS i.c.s to be used.

The best technique is to write the index numbers and letters on to the circuit diagram (as shown in *Figure 1.27*), so that the circuit can then be constructed with reference to these indexes. If i.c.s are always placed in the same way, for example a single 16-pin i.c. with pin 1 on row A1, then construction

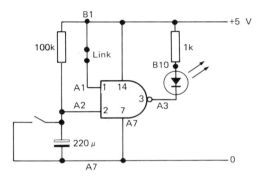

Figure 1.27. A typical circuit using the number/ letter index system of the Eurobreadboard for easy construction. The number/letter references show where i.c. pins or component leadout wires plug into. Any of the five plug-in points along one line of the numbered contact can be used

can be very rapid and circuit changes can be made rapidly. A useful hint, incidentally, is to connect 1 MΩ resistors between all CMOS inputs and earth. If these resistors are left undisturbed, then rearrangements of the circuit can be made without risk to the CMOS i.c. and without having to remove the i.c. When TTL i.c.s are used, inputs which are to be kept at logic 1 can be left floating (unconnected) providing that the i.c.s are operated with low frequency signals.

2 Logic gates

A *logic gate* is a circuit which uses digital signals as its inputs and outputs. What makes a circuit a gate is that each output depends *entirely* on the signals applied at the inputs. If these input signals change, then the output signal may also change. Digital circuits which use logic gates are usually arranged so that a logic 1 appears at an output only for some definite combination of input signals — for this reason these circuits are sometimes called *combinational logic circuits*. In theory, we could make i.c.s for each and every possible combination of input signals to produce a 1 output, but this would be wasteful of resources. In practice, what is done is to make i.c.s which accomplish a few standard logic operations. From these standard logic i.c.s any combinational logic circuit can be built up. The microprocessor is an extension of this idea — a circuit which can perform virtually any logic function.

The action of a standard combinational logic circuit, or of any circuit made up from these units, can be described in two ways. One way is by the use of a *truth table*. A truth table shows what output can be expected from each possible combination of inputs, so that the action of the circuit can be readily checked. Another method of describing the action of a circuit is by *Boolean algebra*. This method is much more concise but less easy for the raw beginner to interpret, so both methods will always be used together in this book.

Boolean algebra, incidentally, was invented long before modern computers. It is named after George Boole (1815—1864) who devised it as a method of turning logical statements into algebraic expressions. Little use was made of this work until Shannon found in 1938 that Boolean algebra could be used to analyse relay circuits which carried out the sort of switching operations we now refer to as 'AND' and 'OR' gates.

AND gate

Figure 2.1 shows the symbol, truth table and Boolean expression for the AND gate, one of the standard logic gates. As the truth table shows, the output is logic 0 unless *both* inputs of the two-input gate are at logic 1. For a three-input AND gate, the output is at logic 1 only when all three inputs are at logic 1. As *Figure 2.2* shows, the truth table for a gate with three

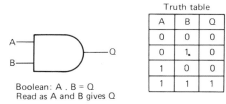

Truth table

A	B	Q
0	0	0
0	1.	0
1	0	0
1	1	1

Boolean: A . B = Q
Read as A and B gives Q

*Figure 2.1. A two-input AND gate symbol,
truth-table and Boolean expression. The output
is 1 only when both inputs are at 1*

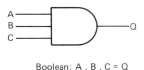

Boolean: A . B . C = Q

Truth table

A	B	C	Q
0	0	0	0
0	0	1	0
0	1	0	0
0	1	1	0
1	0	0	0
1	0	1	0
1	1	0	0
1	1	1	1

*Figure 2.2. A three-input AND gate. The number of lines of the
truth table equals 2^n, when n is the number of gate inputs. The
output is 1 only when all the inputs are at 1, whatever the number
of inputs*

inputs takes up a lot more space than that for the two-input gate, and is prohibitively long when four or more inputs are used. The Boolean expression uses the dot symbol to indicate the AND action so that for a four-input AND gate the action is A.B.C.D = 1. In words, this means that the output is 1 *only* when A *AND* B *AND* C *AND* D inputs are all at 1. Note that the action of the AND gate can also be illustrated by switches in series, as shown in *Figure 2.3*.

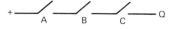

For each switch, open = 0, closed = 1

Figure 2.3. A switch circuit which has the same truth table as a three-input AND gate

AND gates are useful for testing for coincidence of binary 1s, and in circuitry where several factors control an output. For example, a gas central heating boiler can be switched on (1) when the room thermostat, the boiler water thermostat, the gas flame detector and the outside temperature detector are all giving a logic 1 output.

OR gate

The symbol, truth table and Boolean expression for the OR gate are shown in *Figure 2.4*. This type of OR gate produces a logic 1 at the output if either input is at 1 or if both inputs

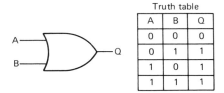

Truth table

A	B	Q
0	0	0
0	1	1
1	0	1
1	1	1

Boolean: A + B = Q
Read as A or B gives Q

Figure 2.4. A two-input OR gate symbol, truth table and Boolean expression. The output is 1 if any input is at 1

are at 1. For inputs which are all set to 0, the output also is
at logic 0. As before, the truth table for an OR gate with
more than three inputs is rather bulky, and the action is most
easily summed up by the Boolean expression $A + B + C + D
+ \ldots = 1$.

The choice of the + sign may look unfortunate at first
sight because it is more natural to associate + with the AND
operation, but the laws of Boolean algebra make the choice of
the + sign automatic. The word 'plus' should not be used when
reading the expression; for +, read OR at all times. The Boolean
expression $A + B + C = 1$ is therefore read as A *OR* B *OR* C
gives logic 1 at the output. Note that the action of the OR
gate can also be illustrated by switches in parallel, as shown
in *Figure 2.5*. An OR gate would be used when more than

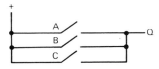

*Figure 2.5. A switch circuit which
has the same truth table as a three-
input OR-gate*

one input may be used to produce an output. On a central
heating system, for example, a logic 1 output may be wanted
(to light the boiler) when the circulating pump starts OR when
hot water is drawn.

Logic NOT

The *NOT gate*, inverter, or complementing circuit, illustrated
in *Figure 2.6*, consists simply of an inverting switch whose
output is the logic opposite of the input. In Boolean symbols,
the action is represented by drawing a bar over a letter which
represents an input. For example, a signal A passed into an
inverter becomes $\overline{A}$, meaning that if A were 0, $\overline{A} = 1$, and for
$A = 1, \overline{A} = 0$.

An inverter may be needed in a circuit if previous gates
have produced an output which is of incorrect polarity, or if
an input is of incorrect polarity. For example, if we need an

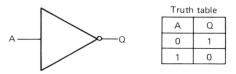

Truth table

A	Q
0	1
1	0

Boolean: $\overline{A}$ = Q
Read as inverse (or complement) of A is Q,
alternatively as Q = NOT A

Figure 2.6. The inverter, complementer, or NOT gate. This simply inverts the signal at its input. The small circle at the output indicates inversion — without the circle the symbol is that of a non-inverting amplifier, or buffer

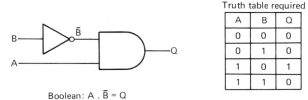

Truth table required

A	B	Q
0	0	0
0	1	0
1	0	1
1	1	0

Boolean: A . $\overline{B}$ = Q

Figure 2.7. Using an inverter to obtain (or implement) a truth table which is not obtained from the AND or OR gate directly

output from a gate when input A = 1 and input B = 0, then one solution is to use the inverter to convert B = 0 to $\overline{B}$ = 1, and then combine with A in an AND gate (*Figure 2.7*).

NAND and NOR gates

The combination of the NOT gate with the AND/OR gates produces gate circuits known as 'NAND' and 'NOR' gates, whose symbols, truth tables and Boolean expressions are illustrated in *Figure 2.8*. These gates might seem of little interest, but they are simpler to construct in integrated form and have the peculiar advantage that they can be converted more easily into other gates. For example, two NAND gates

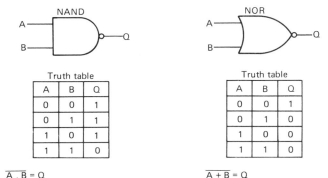

$\overline{A \cdot B} = Q$
Read as NOT (A and B) gives Q

$\overline{A + B} = Q$
Read as NOT (A or B) gives Q

Figure 2.8. The NAND and NOR gates, with truth tables, symbols and Boolean expressions

connected as in *Figure 2.9* form an AND gate; it is not possible to combine two AND gates to form a single NAND gate. For this reason the NAND/NOR gates are much more common

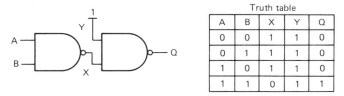

Figure 2.9. Two NAND gates used to simulate the AND gate. The truth table shows how this happens — the second NAND gate acts as an inverter because one input is held at logic 1

and readily available than other types. Further examples of the uses of these gate types in forming other gates are given at the end of this chapter.

Exclusive-OR

The simple OR gate has a logic 1 output for any input or combination of inputs at logic 1. The exclusive-OR (X-OR) gate has its output at logic 1 for any *single* input at 1, but not for a

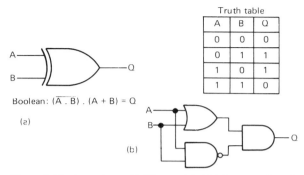

Figure 2.10. A two-input X-OR gate (a). This has a truth table similar to that of the OR gate, except for the state when both inputs are 1. The Boolean expression shows this exclusion by using the NOT (A and B) term. (b) An arrangement of gates which gives X-OR action

combination of inputs. The truth table and symbol are shown in *Figure 2.10* along with the Boolean expression which, in this case, is rather more complex. As it happens, though X-OR gate i.c.s are obtainable, the Boolean expression contains the clue to a method of constructing an X-OR gate from other circuits (see later, this chapter).

Discovering gate action

The usefulness of standard gate actions is that any gating logic which may be needed can be obtained from a suitable combination of these logic circuits. Designing a logic circuit is not entirely straightforward, but by following the rules below discovering the action of a circuit is relatively simple:

(1)   Label with a letter each separate input for all the gates in the circuit, and label the output (Q) also.

(2)   Draw up a column for each lettered input and final output, setting the primary inputs (of the first gates in the circuit) apart from the others.

(3) Write rows of logic 0s and 1s for the primary inputs. Each
row should be a binary number, starting with 000. . . .
(as many 0s as there are primary inputs), and increasing
by 1 in each row. The total number of rows should then
be 2^N, where N is the number of primary inputs. For
example, if N = 3 (three primary inputs), $2^N = 8$, so that
the truth table should have eight rows.

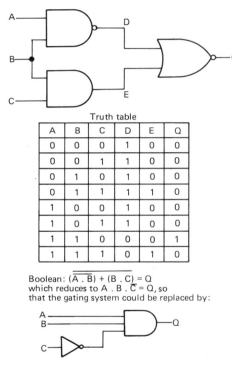

Truth table

A	B	C	D	E	Q
0	0	0	1	0	0
0	0	1	1	0	0
0	1	0	1	0	0
0	1	1	1	1	0
1	0	0	1	0	0
1	0	1	1	0	0
1	1	0	0	0	1
1	1	1	0	1	0

Boolean: $\overline{(\overline{A \cdot B}) + (B \cdot C)} = Q$
which reduces to $A \cdot B \cdot \overline{C} = Q$, so
that the gating system could be replaced by:

*Figure 2.11. Discovering the truth table for a gate
system. The inputs are listed, and the intermediate
signals C and D noted down, knowing the truth
tables of the AND and NAND gates. The final output Q is written down knowing the C and D states
and the truth table for the NOR gate*

(4)   Now write in the output for each primary gate, using the
       truth tables for AND, OR, NOT, NAND, NOR, X-OR
       as required. If these outputs then form the inputs to
       other gates, complete the truth tables for these also until
       the output column is complete. The result is the complete
       truth table for the gating circuit.

Figure 2.11 shows an example of this technique being used
for a circuit with three primary inputs and one output. When a
large number of primary inputs exist, however, the truth table
is very unwieldy, and a better approach, as usual, is to use
Boolean symbols. Each primary input is labelled A, B, C. . . .
as before, but the outputs of the primary gates are then
labelled with the Boolean expressions which describe the gate
action. For example, if a NAND gate has inputs A, B, then the
output is labelled $\overline{A.B}$ (NOT A-and-B). The next set of gates
is labelled in the same way, using the inputs which have already
been discovered.

The final expression represents the gate action but may
require simplification to remove terms which are surplus to
needs. The laws of Boolean algebra shown in *Table 2.1* can be
used to simplify such expressions, and one example is shown in
Figure 2.12. In general, though, these methods are required
only by the professional engineer. The laws are shown only
for reference purposes so that the mathematically inclined
reader may make use of them.

The opposite process can usually best be carried out by the
use of Boolean algebra. Once the Boolean expression for the
action of a circuit has been found, the circuit can be sketched
out — but finding the Boolean expression may be a problem.
To illustrate how a circuit can be constructed (or implemented)
using the Boolean expression, consider the expression for the
X-OR gate of *Figure 2.10*:

$$(\overline{A \ . \ B}) \ . \ (A + B) = Q$$

(A + B) indicates that an OR gate is needed with A and B as
inputs. $(\overline{A.B})$ indicates that a NAND gate is needed, also with
A and B as inputs. Finally the outputs of these gates must be

Table 2.1

Gate laws:

AND	OR	NOT
$0 . 0 = 0$	$0 + 0 = 0$	$\overline{1} = 0$
$0 . 1 = 0$	$0 + 1 = 1$	$\overline{0} = 1$
$1 . 0 = 0$	$1 + 0 = 1$	
$1 . 1 = 1$	$1 + 1 = 1$	

Boolean algebra laws:

A, B are logic quantities which may be 1 or 0

Gating:	$A + 0 = A$
	$A + 1 = 1$
	$A . 0 = 0$
	$A . 1 = A$
Identity:	$A + A = A$
	$A . A = A$
Order:	$A + B = B + A$
	$A . B = B . A$
Change-round:	$(A + B) + C = A + (B + C)$
	$(A . B) . C = A . (B . C)$
Factorising:	$(A . B) + (A . C) = A . (B + C)$
	$(A + B) . (A + C) = A + (B . C)$
Complements:	$A + \overline{A} = 1$
	$A . \overline{A} = 0$
Double negative	$\overline{\overline{A}} = A$
De Morgan's Theorem:	$\overline{A . B} = \overline{A} + \overline{B}$
	$\overline{A + B} = \overline{A} . \overline{B}$

Note: The names shown above are not necessarily those used by mathematicians

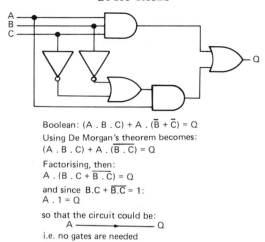

Boolean: (A . B . C) + A . ($\overline{B}$ + $\overline{C}$) = Q

Using De Morgan's theorem becomes:
(A . B . C) + A . ($\overline{B . C}$) = Q

Factorising, then:
A . (B . C + $\overline{B . C}$) = Q

and since B.C + $\overline{B.C}$ = 1:
A . 1 = Q

so that the circuit could be:
   A ——————————→ Q

i.e. no gates are needed

*Figure 2.12. Showing how the use of Boolean
algebra can simplify a logical expression and avoid
the use of unnecessary gates*

combined in an AND gate, as indicated by the AND dot between the sets of brackets. The circuit is therefore as indicated
in *Figure 2.10(b)*.

This comparatively simple example could also have been
drawn out using only the truth table. The truth table shows
that a 1 output must be obtained when the inputs are OR-ed
but not when both inputs are 1. A NAND gate produces a 0 at
the output when both inputs are 1, so that an AND connection
of the NAND gate with an OR gate should give the required expression. Once a gating arrangement has been designed in this
way, it should be checked by constructing a truth table as
shown earlier, unless the number of inputs is prohibitively
large.

A gating system may have to be constructed, however, with
only verbal information. For example, the specification may
be: a 1 output is required when either of inputs A and B is 1,
but with inputs C and D 0. No other combination must give
a 1 output. This may be drawn up as a truth table, but with

four inputs 16 rows (2^4) of inputs will be needed. The alternative is to construct a Boolean expression directly from the specification. It is fairly obvious that a 1 at the output for either A *or* B inputs can be obtained by using an OR gate, but the output of this gate must produce a final 1 output only if neither C *nor* D produces a 1. Now these two inputs can be fed to a NOR gate, whose output will be 1 only while C and D are both 0. By combining the output of this NOR gate with the output of the OR gate, using this time an AND gate, the correct function should be obtained. In Boolean terms the verbal expression would be written:

$$Q = (A + B) \cdot \overline{C} \cdot \overline{D}$$

By De Morgan's Theorem (*Table 2.1*),

$$\overline{C} \cdot \overline{D} \text{ converts to } (\overline{C + D})$$

so that the expression becomes

$$Q = (A + B) \cdot (\overline{C + D})$$

which is the combination of OR and NOR AND-ed as shown above.

One of the fascinations of logic circuit design is that it cannot entirely be reduced to rules; intuition and experience still have a part to play. Very often an experienced designer will be able to avoid difficulties which are not at all obvious to another designer who is laboriously ploughing through the set procedure.

Race hazards

Logic systems using signals which change only at widely separated intervals are seldom troubled by problems arising from the time delay in each gate. Some logic systems, however, must deal with signals which are at the 1 or 0 levels only for short times of the order of one microsecond or less. Such high-speed logic presents difficulties arising from the time delays in different circuits which are called *race hazards*.

Imagine, for example, a NAND gate which generates a negative-going pulse when its inputs both go to logic 1. The duration of this negative-going pulse will be the time for which both inputs are positive. Imagine now that both inputs are pulses 1 μs long. If they coincide exactly then the output pulse will be 1 μs long, but if one input pulse is delayed by 300 ns (1000 ns = 1 μs), then the duration of the output pulse can only be 700 ns. If one input pulse is delayed by 1 μs or more, then there will be no change in the output at all. Designing a circuit to eliminate race hazards is not a task for the beginner, and the main reason for mentioning the problem is to show why logic circuits which work perfectly at low speeds may fail in curious ways when high pulse rates are used.

Arithmetic circuits

As the name suggests, *arithmetic circuits* are intended to carry out the functions of binary addition, subtraction, multiplication and division. Thanks to the use of 2's complement arithmetic, subtraction can be carried out by using a NOT gate and an adding circuit, and the actions of multiplication and division can also be carried out using only an adder arithmetic circuit combined with the use of shift registers (Chapter 7).

The addition of two binary numbers on paper is straightforward, but the arrangement of logic gates to perform the same action is less simple. The least-significant bits are easiest to add because there is no carry figure from an earlier stage. Adding a 1 to a 0 will produce a 1 in the output, but adding

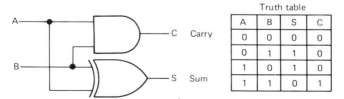

Truth table

A	B	S	C
0	0	0	0
0	1	1	0
1	0	1	0
1	1	0	1

Figure 2.13. A half-adder circuit and its truth table

1 to 1 will produce a 0 at the main output and a carry figure of 1 to be added to the next pair of bits. The truth table for this section, called the *half-adder*, is therefore as shown in *Figure 2.13*. The output in the sum column is the same as would be expected from an X-OR gate, while the carry output is the result of an AND operation on the input parts. A simple half-adder can therefore be constructed using the gate circuit shown in *Figure 2.13*. Half-adders can also, of course, be obtained as MSI integrated circuits or as a small part of LSI circuits.

For adding all the other pairs of bits a more elaborate circuit called the *full-adder* must be used. The reason is that three inputs are now needed, two for the bits which are being added and another one for the carry bit which has come from the previous addition. The truth table is now as shown in *Figure 2.14*; check this with an addition if there is any uncertainty.

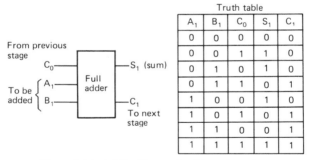

Truth table

A_1	B_1	C_0	S_1	C_1
0	0	0	0	0
0	0	1	1	0
0	1	0	1	0
0	1	1	0	1
1	0	0	1	0
1	0	1	0	1
1	1	0	0	1
1	1	1	1	1

Figure 2.14. A full-adder truth table and symbol

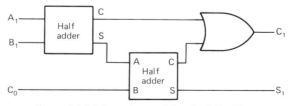

Figure 2.15. The gating used for the full-adder

The full-adder is obtainable as a complete MSI i.c., but it can be constructed from standard gates as shown in *Figure 2.15*. The reader should check, using the procedure noted earlier, that this set of gates will produce the correct truth table.

Negative signs

In discussing arithmetic so far, it has always been assumed that the numbers are positive. For many digital applications, negative numbers have to be manipulated, and some way of indicating the sign of such a number is needed. An obvious method of writing a negative number is to use the 2's complement of a positive number so that, for example, the number 100101 (decimal 37) converts to 011011 (decimal 27). Adding this to a six-bit binary number produces the same effect as the subtraction of 37; for example, adding to 111001 (decimal 57) gives 010100 (decimal 20) when the seventh bit is discarded as shown in *Figure 2.16*.

```
  1 0 0 1 0 1 ─────────► 0 1 1 0 1 1        2's complement
  decimal 37
                         + 1 1 1 0 0 1
                         ────────────
                    (1) 0 1 0 1 0 0
```

Answer: 0 1 0 1 0 0   after discarding carry

Figure 2.16. Two's complement arithmetic

This is logical enough but the problem is that a digital circuit cannot distinguish between 011011 meaning decimal 27 and 011011 meaning decimal −37; and a readout of such binary numbers could not distinguish between the positive number and the negative one. This is done by using numbers in the eight-bit byte form, and reserving the most significant bit for indicating sign. The other seven bits are used for representing number, so that when this method of writing a signed binary number is used only seven-bit numbers can be used in one byte. If we want to use larger numbers, we can have each number represented by a double byte of 16 bits, with the most significant bit (the 16th) representing the sign. That then leaves 15 bits to represent numbers, giving a much greater range. Similarly three-byte or four-byte numbers can be written as signed binary numbers with the most significant bit reserved for indicating the sign.

The convention as to what value of the most significant bit shall represent a negative number is easily decided. When, in our earlier example, decimal 37 is written as an eight-bit byte, it becomes 00100101. Converting this to − 37 by taking the 2's complement gives 11011011, so that the most significant bit is 1, indicating a negative value. The number 01011011 would be a positive number (decimal +91). A 0 as the most significant bit therefore indicates a positive number, and a 1 indicates a negative number when this convention is used.

The advantage of this scheme is that arithmetic can be carried out on these bytes without any regard to whether the binary numbers represent positive or negative numbers, and it is only when a final result is obtained that a 1 in the most significant bit has to be interpreted either as 256 (if only positive numbers are used) or a negative sign if signed numbers are used.

If signed binary numbers are in use, a 0 as the most significant bit indicates that the following digits represent a positive number, and the conversion of binary to decimal is carried out in the usual way. If the most significant bit is 1, however, a negative sign is written, and the following bits are taken as a 2's complement number. To find the decimal equivalent, therefore, 1 is subtracted from the least significant bit and the number is complemented. For example, the number 11011011 is read in the following stages:

(a)　　the MSB indicates a negative sign;
(b)　　1011011 has 1 subtracted, giving 1011010;
(c)　　this is complemented to 0100101;
(d)　　the decimal equivalent of 0100101 is 37, so that the number represented by 11011011 is −37.

Latching

A *combinational logic circuit* produces an output which is completely determined by the inputs, so that a given combination of inputs always produces the same output. Logic gate circuits can also be combined to produce circuits, called *sequential logic circuits*, which behave quite differently.

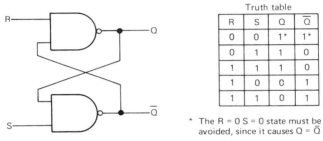

Truth table

R	S	Q	Q̄
0	0	1*	1*
0	1	1	0
1	1	1	0
1	0	0	1
1	1	0	1

* The R = 0 S = 0 state must be avoided, since it causes Q = Q̄

Figure 2.17. The R−S flip-flop

The circuit of *Figure 2.17* is an example of such a sequential logic circuit. The circuit uses two NAND gates, cross coupled as shown, with two inputs and two outputs. At first sight, the truth table seems quite normal, but repeating some of the inputs demonstrates that the output depends on the *sequence* of inputs and not simply on the combination.

For example, with inputs S = 0, R = 1 the output is Q = 0, Q̄ = 1 and changing the inputs to S = 1, R = 1 causes no change. For inputs S = 1, R = 0 the output is Q = 1, Q̄ = 0, and changing R = 0 to R = 1, so that S = 1, R = 1, again causes no change. For this circuit, therefore, the inputs S = 1, R = 1 can produce either a 1 or a 0 at one output, depending on what the immediately previous inputs were.

The circuit of *Figure 2.17* is called an R−S *latch* or *flip-flop*; the word 'latch' suggests that the output is locked in place. Latch circuits are used to retain an output for some time even when the inputs are changing, and such circuits are dealt with more fully in Chapter 4. The simple R−S latch has few applications.

Practical work

Practical work on combinational logic circuits can be carried out using the breadboard along with a few standard i.c.s. In the 74 TTL range, the 7400 quad NAND gate is particularly useful; its pin diagram is shown in *Figure 2.18*. Logic circuits

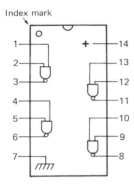

Figure 2.18. The pinout diagram for the SN7400N quad NAND gate

can be connected up using switches as inputs, remembering that an open circuit input on a TTL chip will be at logic 1. Output states can be detected using l.e.d.s with their series current limiting resistors. If the CMOS equivalent (CD4011A) is used, remember that each input should have a 1M resistor permanently connected to earth, and that the output is best connected to a transistor/l.e.d. combination as indicated in *Figure 1.25*.

The circuit arrangements of *Figure 2.19* simulate standard gate actions for which the truth table can be found either practically or by using the methods of analysis mentioned earlier.

Figure 2.20 is the internal circuit of the 7482 two-bit full-adder, with truth table. The carry bit between stages is internally connected and is not available at the pins. The

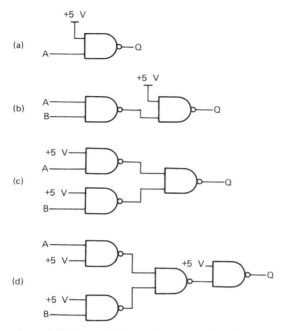

Figure 2.19. A practical exercise – connecting the gates of a 7400 so as to simulate the actions of other gates

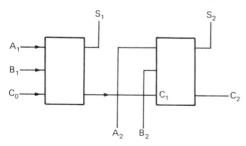

Carry C_0 added to A_1 added to $B_1 \longrightarrow S_1$ and carry C_1
Carry C_1 added to A_2 added to $B_2 \longrightarrow S_2$ and carry C_2

Figure 2.20. The arrangement of adders inside the 7482

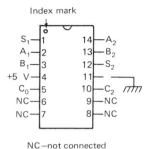

Index mark

NC—not connected

*Figure 2.21. Pinout diagram for the
7482 dual adder*

pinout of the i.c. (*Figure 2.21*) shows what connections have to be made to switches to give suitably binary inputs, using A_1, A_2 as one pair of bits and B_1, B_2 as the higher order.

Since the circuit is of a full-adder, a carry into the least significant bits (C_0) can be added, and the carry out (C_2) from the most significant bits can be detected.

3 Clocks and other inputs

Simple combinational or latching logic circuits operate immediately on their inputs, involving no time other than the small delay caused by each gate. A clocked circuit introduces a completely new idea to logic circuits — the idea that all operations take place during the short time of a voltage pulse called the *clock pulse*. The use of a clock pulse is essential to sequential logic circuits because the clock pulse sets the time between the steps of a sequence.

A typical clock pulse is of the form illustrated in *Figure 3.1*. Between clock pulses the voltage level is logic 0, and at

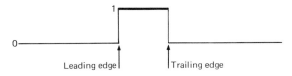

Figure 3.1. A typical clock pulse

the leading edge of the clock pulse the voltage rises abruptly to logic 1. The time needed for this voltage rise is called the *rise time*, and will be a time of several nanoseconds. The voltage remains at logic 1 for a time called the *pulse width*, and then returns to logic 0 in a time called the *fall time*. Circuits which make use of clock pulses will operate on either the leading edge or the trailing edge of the clock pulse.

Figure 3.2 is of a clocked R—S latch. The truth table for this circuit shows the output values before and after the clock pulse. The important difference between this design and the simple R–S latch is that the output does not change until the clock pulse arrives, and then the output remains unchanged (latched) until the next clock pulse arrives, assuming that the input has not changed in that time.

A clocked latch or flip-flop such as this example will act as a temporary store for a binary digit, and it is particularly useful when a fast-changing output, such as that of a calculator,

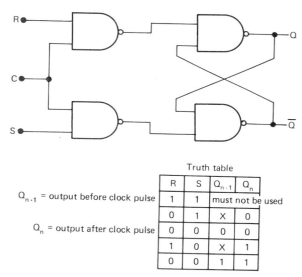

Truth table

		R	S	Q_{n-1}	Q_n
		1	1	must not be used	
Q_{n-1} = output before clock pulse		1	1		
		0	1	X	0
Q_n = output after clock pulse		0	0	0	0
		1	0	X	1
		0	0	1	1

Figure 3.2. A clocked R–S flip-flop. While the clock input is low the inputs to the flip-flop gates are both high, keeping the outputs in the state to which they were previously set. Changes at the R and S inputs have no effect until the clock input goes high. In the truth table, X means 'don't care' — the input may be either 1 or 0

has to feed information to a slow-operating device such as a printer. By using the slow device to generate a clock pulse whenever it is ready for more information, the latch can be used to hold information until it is needed. In microprocessor circuits eight-bit latches are very often used for such a purpose.

Clocking is essential for LSI circuits because signals reaching a point in the circuit may have come through different numbers of gates thereby causing different delays. By holding back the action until a clock pulse arrives, and making sure that the time between clock pulses is always more than the greatest possible delay that a signal can suffer, all the problems caused by signal delays disappear.

The shape of clock pulses is important. The circuits to which clock pulses may be applied include high-gain amplifiers which are active as amplifiers only during the time when they are switching over between logic 0 and logic 1. If this time is not very short (around 35 ns) these amplifier circuits may oscillate during the changeover, thus generating a few clock pulses of their own. These false pulses can, in turn, be carried around the circuit causing incorrect operation.

Many clocked circuits specify maximum rise and fall times for the clock pulses, and correct operation cannot be guaranteed if these times are exceeded. The time that is specified is that taken for the voltage to rise or fall between the 10% and 90% levels. For a system using a 5 V signal, for example, the rise time would be the time taken for the voltage to change from 10% of 5.0 V (= 0.5 V) to 90% of 5.0 V (= 4.5 V), and the fall time would, of course, be the time needed for the voltage to fall from 4.5 V to 0.5 V.

Another cause of false pulses is the mechanical switch. When mechanical switches are used to test gates, as was done in the previous chapter, no problems arise because the final result is read from the l.e.d.s. When a switch is used to generate clock pulses, however, or any pulses which are to be counted, then the problem of *contact bounce* arises.

The materials of a mechanical switch are springy, so that the controls will bounce a few times when closed by a mechanical switch thereby causing several pulses, one at each bounce, and if the switch is used to generate clock pulses then each bounce pulse will count as a clock pulse.

Wherever a mechanical switch is used, therefore, it must be connected to a circuit called a *debouncing circuit* which removes the additional pulses caused by the bouncing contacts.

One such circuit, as shown in *Figure 3.3*, uses a single-pole two-way switch together with two NAND gates arranged as an R—S latch. The action is as follows:

With the switch in the 1 position, the R input of the R—S latch is connected to logic 0, and the S input is at logic 1. This produces a 1 output at Q. When the switch is changed over, the first part of the action of the switch causes both R and S

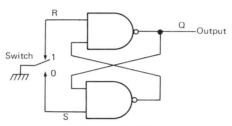

Figure 3.3. Using an R—S flip-flop circuit to 'debounce' a switch. This is called a 'hardware debounce'. In microprocessor circuits debouncing is carried out by programming a time delay before the switch value is read — this is a software debounce

contacts to be broken, so that both inputs will be at logic 1. From the truth table for the R—S latch (*Figure 2.17*) this does not affect the output, and there is no effect either if the switch leaf bounces back against the 1 contact. When the switch leaf first makes contact with the 0 connection, the effect is to connect the S input to logic 0. This immediately switches the latch to output 0, because the R input is at logic 1. If the switch bounces open again, the effect is simply to allow both R and S input to rise to logic 1, keeping the output unchanged.

Another type of debouncing circuit is shown in *Figure 3.4*. This method makes use of a *Schmitt input circuit* which can be obtained associated with NAND gates (7413, 74132) or inverters (7414). A Schmitt circuit has what is called *backlash*, that is it will switch in either direction but not at the same voltages. For example, a Schmitt input may switch over in one direction at +4.0 V, but will not switch back until +1.0 V has been reached. The speed of the switchover is governed entirely by the circuits inside the device, so that a very fast rise or fall of voltage will be generated no matter how slowly the input voltage changes.

In the circuit of *Figure 3.4*, the switch is a single pole on/off type and it is connected to a capacitor and resistor as shown;

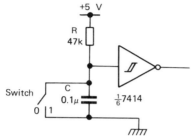

Figure 3.4. Another type of hardware debounce using a Schmitt-trigger gate, one of the six Schmitt inverters in the 7414 package. The symbol shown inside the inverter symbol indicates Schmitt action

With the switch open, the connection of R to +5.0 V ensures that the input of the inverter or gate is logic 1, so that the output is at logic 0. When the switch closes, the first touch of the switch contacts discharges C, so that the input voltage drops to logic 0, making the inverter output switch over to logic 1. If the switch contacts now bounce open again, the input of the inverter cannot *immediately* rise to a voltage which will cause the output to switch, because of the time needed for capacitor C to change. The values of R and C are chosen so that the capacitor cannot charge to a voltage sufficient to change over the inverter in the time of one switch-bounce.

Schmitt-trigger input circuits, which are available both in TTL and in CMOS types, are extremely useful when clock pulses are obtained from transistor oscillators or other circuits which do not use digital circuitry. Any shape of wave, for example, whose amplitude is limited to 5.0 V, will switch a

Schmitt circuit so as to produce clock pulses whose rise and fall times are suitable for operating other TTL or CMOS circuits. Fast rise times are not as important for the correct operation of CMOS circuits as they are for TTL circuits, but better shaped clock pulses are always an advantage. The inputs may be rectified a.c., sinewaves, or any other waveshape provided that the amplitude does not exceed the maximum input amplitude permitted for the circuit.

If the signals used to convert to clock pulses are obtained from the same supply voltage as that used for the digital circuits, no problem of limiting amplitude will arise. Input signals of various types, however, often come from quite different types of circuits operating at different voltages, and

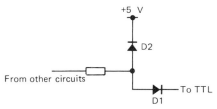

Figure 3.5. An interface circuit which limits the amplitude of a pulse, obtained from other circuits, to the 5.0 V needed for TTL use

are capable of damaging the inputs of digital circuits if applied directly. These inputs need not necessarily be clock inputs — any input to a digital circuit will have to be limited so as not to exceed the normal 1 and 0 limits for the type of circuit that is used.

One method of limiting amplitude is shown in *Figure 3.5* using diodes. Diode D1 will conduct only when the signal voltage is above 0.5 V, so that the input of the digital circuit is protected against negative signal inputs. Diode D2 will conduct if the signal voltage exceeds 5.5 V, so that the voltage applied to the diode D1 cannot exceed 5.5 V. This in turn ensures that the input to the digital circuit cannot exceed 5.0 V, because of the 0.5 V drop across a conducting diode.

Another method uses a transistor as a voltage limiter. With the collector of the transistor fed from a +5.0 V supply, the voltage at the collector must swing between +5.0 V (transistor cut-off) and 0.2 V (transistor fully conducting). A resistor in series with the base input prevents excessive driving currents,

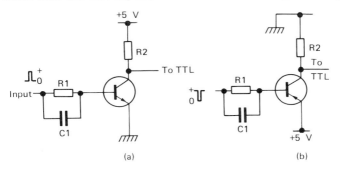

Figure 3.6. A transistor interface: (a) n-p-n, (b) p-n-p

and a small capacitance in parallel with this input resistor prevents the edges of pulses from becoming excessively rounded by the capacitance between the base and the emitter. The circuit can be used in two forms: on the *n-p-n* form a sharp negative leading edge is created; the *p-n-p* form creates a sharp positive leading edge.

Generating clock pulses

Clock pulses for circuits which are clocked at high speed usually have to be generated, unless pulses from other sources (such as the mains supply) can be used. A clock pulse generator circuit can make use of transistors, digital or linear i.c.s, provided that the same power supplies can be used and that the pulse will be of a suitable rise and fall time. A linear i.c. circuit which is very often used along with digital circuits as a source of clock pulses is the 555 timer. *Figure 3.7* shows a 555 timer circuit which generates clock pulses whose frequency

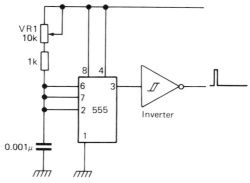

Figure 3.7. A clock pulse generator using the 555 timer. The output from the timer i.c. is inverted so that the short flyback pulse of the timer is used as a clock pulse

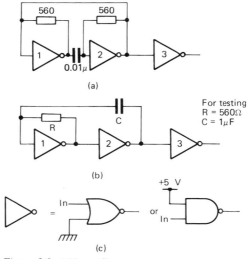

Figure 3.8. TTL oscillator circuits. The i.c.s shown as inverters may also be NOR or NAND gates connected as indicated. The circuit shown in (b) is much more common

can be varied by changing the setting of VR1. The clock pulses produced by this circuit are sharp enough to be suitable for CMOS digital circuits, but a TTL buffer circuit is needed if TTL circuits are to be driven. Using a Schmitt buffer, such as one part of the Hex Inverter 7414, will ensure that the output pulses have short rise and fall times and are able to drive 10 input circuits.

An alternative method of obtaining clock pulses is to use a digital oscillator circuit. Oscillators can be made using NAND or NOR gates, using circuits such as those shown in *Figure 3.8*. Note that the TTL circuits need low resistor values, with 1 kΩ as the maximum value.

This is because the input to a TTL circuit is to the emitter of a transistor whose base is connected to +5.0 V. A high resistance at the input will prevent this emitter from reaching the low logic 0 voltage, so that normal switching is impossible. The current drawn at logic 0 is 1.6 mA so that the 1 k in series with a TTL input causes the voltage at the input to rise to +1.6 V, rather high to be reliably taken as logic 0.

In the circuit of *Figure 3.8(b)* the connection of resistor R between the output and the input of gate 1 creates negative feedback, which reduces the gain of the gate and causes a wider range of linear operation. Imagine that the voltage at the input of gate 1 is low. The output of gate 1 will be high and because gate 2 is also an inverter, its output will be low. Because of the connections of R, capacitor C will start to charge. One plate is held at low voltage by the output of gate 2, but the other plate can be charged by current through R. As the voltage at the input of gate 1 rises, the inverting action will cause the output voltage of gate 1 to drop, switching gate 2 so that its output is high. The voltage pulse through the capacitor will cause the input voltage of gate 1 to rise sharply to logic 1, and the capacitor will now discharge through R because the output of gate 1 is now at logic 0. The times for which the gate voltages are steady are not equal because when the input of gate 1 is at logic 1 the only resistance in circuit is R. When the input of gate 1 is at logic 0, however, the input resistance of the gate (which varies from one i.c. to another) is in parallel with R.

The third gate in the circuit sharpens up the rise and fall times of the wave. Note that when NAND gates are used, unwanted inputs should be connected to logic 1; when NOR gates are used these other inputs should be connected to logic 0.

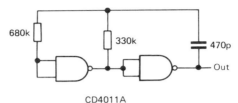

CD4011A

Figure 3.9. A typical CMOS oscillator – NAND gates have been shown, but NOR gates could also be used. The CD4069 hex inverter is also suitable

The CMOS circuits can use much higher resistance values. TTL oscillators have a reputation for not always starting when the circuit is switched on, and oscillators based on the Schmitt trigger digital gates are preferable, as well as needing fewer sections of an i.c. A typical Schmitt trigger oscillator is illustrated in *Figure 3.10*. Imagine that the input voltage is low. Because the gate is an inverting amplifier the output will be high, at logic 1. Current flowing through R will charge C, so that the voltage across C will rise, following the usual capacitor charging curve with time constant RC. When the input voltage reaches the upper trigger voltage of the Schmitt circuit, the gate switches over and the output voltage goes to logic 0. The

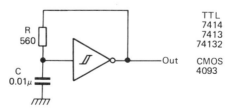

Figure 3.10. A Schmitt oscillator circuit. TTL circuits can use only small resistance values, less than 1k, in this circuit, but CMOS i.c.s enable much higher resistance values to be used

capacitor now discharges through R but the decrease of voltage at the input does not affect the output of the gate until the lower trigger voltage is reached. At this point, the gate switches over again and the action repeats. The rise and fall times are very short because they are controlled by the internal circuits of the Schmitt gate, and the clock pulses are of excellent quality.

The output of the circuit of *Figure 3.10* is a square wave with a mark/space ratio of approximately 1:1. If very short duration clock pulses are wanted, a very useful source is a *unijunction oscillator*. The circuit is shown in *Figure 3.11*. The

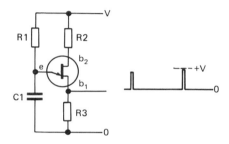

Typical values
R1 = 100k (must be high)
C1 = 10n
R2 = R3 = 100

Figure 3.11. A unijunction oscillator. The output is of very short pulses with fast rise and fall times

unijunction is a trigger device which has two connections to a strip of semiconductor, and one to an emitter junction. When a voltage is placed across the base contacts only a very small current flows until the voltage of the junction is brought up to a level which is a constant fraction of the voltage across the base connections.

When this 'triggering' voltage is reached, the unijunction conducts freely from emitter to b_1 and from b_2 to b_1 so that large transient currents can flow. Limiting resistors must be connected to ensure that these currents are not so large as to damage the unijunction.

In the circuit of *Figure 3.11*, with the capacitor C1 discharged, the emitter voltage is low, so the unijunction is not conducting. As C1 charges through R1, the emitter voltage rises until it reaches the triggering level. At this voltage, the unijunction becomes a good conductor and C1 discharges to earth through the emitter-base junction and through R3, with current flowing through R2 also. The discharge is rapid because of the small value of R3, and a pulse of short duration is generated. The discharge of the capacitor also has the effect of switching the unijunction back to its non-conducting state, so that the process can start again. The pulse from a unijunction is sharp enough for CMOS use and is usually sufficient for TTL clocking also, but in order to drive a number of TTL circuits a buffer is desirable, and a Schmitt inverter such as the 7414 provides both buffering and any sharpening of the waveform which may be needed. The pulse frequency is controlled by the time constant C1×R1.

Keyboard inputs

A series of pulses applied to a circuit is a type of input signal called a *serial input*. The other type of signal input is the *parallel input* which consists of logic signals (either steady voltages or pulses) on a number of input lines. Parallel inputs are very often used in conjunction with keyboards, and one simple example of a keyboard action is a decimal keyboard as might be used in a calculator.

Since digital circuits operate only with binary number inputs, any decimal figure selected by pressing a key must be converted into a binary number. The output of a keyboard which carries out this conversion is said to be BCD (Binary Coded Decimal) because each decimal figure is separately coded as a binary number.

This is *not* the same as a binary *number* which represents the decimal *number*. For example, if we are coding the decimal number 35, pressing the figure 3 key gives binary 0011, then pressing the 5 key gives 0101, so that 35 in BCD is 0011 0101. Each group of four bits represents one figure of the decimal number. In normal binary coding, 35 is represented by 100011, a six-bit binary number. For calculating purposes, if we want to make use of the comparatively simple binary arithmetic, the BCD signals will have to be converted into binary — this can be done using a converter i.c. such as the 74184.

On the keyboard itself, the usual action is for each key to connect a common line to an output line. The coding to BCD

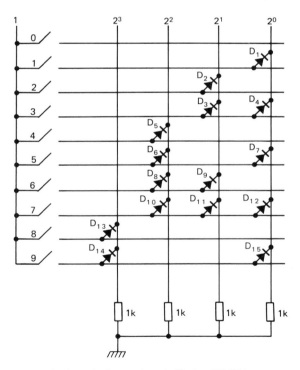

Figure 3.12. A diode matrix suitable for CMOS inputs

can be carried out very simply using a circuit called a diode matrix, shown in *Figure 3.12*. The principle is as follows:

Each binary line is connected through a low value resistor to logic 0, so that with no input keys depressed all lines are at logic 0. When the key marked as decimal 1 is depressed, diode D1 is connected between the positive supply line and the lowest binary line (2^0). Similarly, depressing key 2 will connect the positive line to diode D2, making the 2^1 binary line voltage rise to logic 1. Key 3 connects two diodes D3 and D4 to both of the binary lines 2^0 and 2^1, so giving the correct binary output 0011. Similarly, each key connection raises the

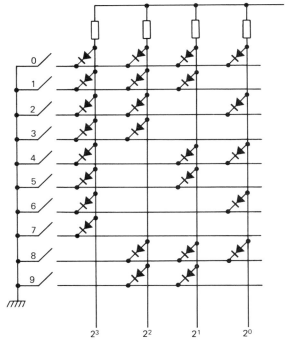

Figure 3.13. A diode matrix suitable for TTL inputs. Several modern keyboards cannot be used this way, and must be connected to a keyboard encoding i.c.

voltages of the appropriate binary lines by conduction through the diodes. This particular type of diode matrix is simple and uses only fifteen diodes, but takes rather a lot of current when used with TTL circuits because of the need to have low resistances in each binary line. CMOS inputs can use resistors of 100 k or more, so that low current operation can be achieved. The diode matrix of *Figure 3.13* is more suitable for use with TTL circuits. In this type of matrix, which uses 25 diodes, each decimal switch causes diodes to be connected to logic 0 so that the TTL input current flows through the diodes. In this type of circuit, however, the output voltage is 1111 when no switch contacts are made, so the digital circuits must be capable of identifying this number (decimal 15) as a 'no-input' signal and gating it out of any counting circuits.

Analogue–digital conversion

Imagine a simple thermistor temperature detector (*Figure 3.14*). The output of this circuit is a voltage which changes as the temperature changes. This type of signal can be displayed

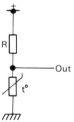

Figure 3.14. A simple temperature detector using a thermistor to give an analogue signal

simply by using a meter, but if we are to obtain a digital readout we need to convert the varying voltage signal (which is an analogue signal) into a set of pulses which can be counted (a digital signal). This requires a circuit called an *analogue-to-digital* (A–D) converter, whose input will be the analogue signal, and whose output will be a series of pulses repeated at intervals. The A–D converter needs, in addition to the analogue signal input, a continuously running clock input which will provide the pulses that form the output, and a reset signal which starts the conversion all over again at intervals.

One very common form of A−D converter is known as a *ramp* type; one form of this type of converter is shown in block diagram form in *Figure 3.15*. A start pulse triggers off a

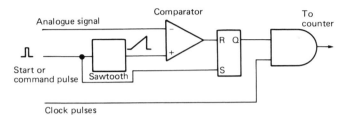

Clock pulses

Figure 3.15. A block diagram of an A-to-D converter, using a ramp conversion method. The start, or command, pulse is at a much lower frequency than the clock pulses, because the ratio of $\dfrac{clock\ pulse\ frequency}{command\ pulse\ frequency}$ is equal to the highest count number that can be obtained. Note that the R−S flip-flop can be used with R = 0 S = 0 because only the Q output is used

linear sawtooth or ramp signal, and also sets the R−S flip-flop because of the connection to the S terminal, so that the Q output of the flip-flop is high. The logic 1 at the Q output of the R−S flip-flop is applied to the AND gate, so that clock pulses are fed into the counter which has been reset to 0 by the start (or *command*) pulse. While the level of the ramp voltage is lower than that of the analogue input voltage, counting continues, but when the two voltages, ramp and input, are equal the comparator resets the R−S flip-flop, so that the Q output of the flip-flop returns to logic 0. This shuts off the AND gate, so stopping the count. The number that has been counted is now proportional to the analogue voltage. The next start pulse then resets the counter and starts another ramp voltage. The start pulses can be derived from the clock pulses or through a delay circuit operated by the output of the flip-flop.

The reverse process is also needed. A *digital-to-analogue* (D−A) converter will convert a series of pulses into a voltage level, counting the pulses and using the number of pulses to control the output voltage. To give one example of the use of such a circuit, a waveform of any shape can be synthesised by a digital circuit, using the digital output to drive a D−A converter. The advantage of this type of signal generation as compared to a conventional analogue type is that any waveform can be generated equally simply. If the digital circuit contains a microprocessor, for example, the waveform of the output can be controlled by the instruction program for the microprocessor, with no alterations needed to the circuit connections.

The principle of most D−A converters is the *resistor ladder network*, a typical example of which is shown in *Figure 3.16*.

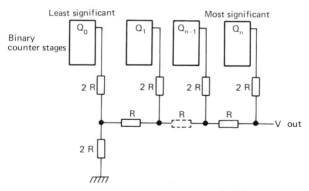

Figure 3.16. A D-to-A converter using a simple ladder network

Each counter has a resistor connected in its output and taken to the series chain or ladder, as shown. The resistor values are not particularly important provided that the *ratios* are correct, so that the resistors are labelled in *Figure 3.16* as R or 2R — one set of resistors has exactly twice the resistance value of the other set. These resistance values will generally be of the order of 1 k, 2 k for TTL circuits, 10 k, 20 k for CMOS. The least significant bit of the counter is connected to the end of the ladder network most remote from the output, and the most significant bit of the counter is connected to the end of the ladder network nearest the output. With this arrangement, the output voltage will automatically be proportional to the binary number output by the counter stages.

The reasons for using this network are that only two values of resistance need to be used, and that the resistance to earth at each point where a counter is connected is constant when all the counters are at logic 0, equal to 2R.

If, for example, the most significant digit output is 1 and all the others are 0, the output voltage must be exactly half of the logic 1 voltage, since the counter feeds through resistance 2R into a load resistance of 2R. When only the next most significant bit is at 1, the output is then one quarter of the logic 1 voltage, as can be seen by the arrangement of resistors redrawn in *Figure 3.17*. Similarly a 1 at the next digit produces

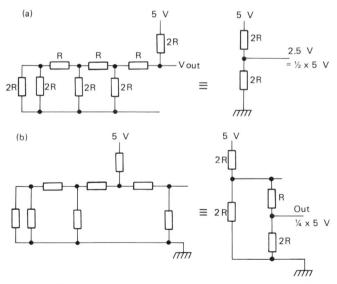

Figure 3.17. Examples of the ladder network: (a) most significant digit 1, others 0, (b) second most significant digit 1, others 0

$\frac{1}{8}$ of the logic 1 voltage and so on. Any binary number at the counter outputs will therefore produce an output voltage in the ladder network proportional to the number counted.

Practical work

The effect of switch bounce is difficult to demonstrate at this stage, but an R–S debouncing circuit (*Figure 3.3*) should be made up. This can be tested by using a crude switch made up from three pieces of wire. The effect of moving the central wire from one of the end wires to the other, and then 'bouncing' it, can be noted − if the debouncing circuit is working correctly then bouncing should have no effect on the output.

Without the use of an oscilloscope the waveforms from digital oscillators and other clock pulse generators cannot be examined, but the output signals can be used to drive an amplifier to show that the circuit is oscillating. The oscillator circuits of *Figure 3.8* should be made up, using the values shown. These values will result in signals which are at audio frequency so that an audio amplifier/loudspeaker combination will produce a sound when fed with such signals.

4 Flip-flops and counters

The R−S latch is a simple form of sequential circuit, but one which has few practical uses. A much more useful type is the edge-triggered D-type flip-flop, which is represented in a diagram by the symbol of *Figure 4.1*. The internal circuit

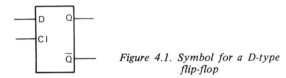

Figure 4.1. Symbol for a D-type flip-flop

consists of gates coupled to each other, but all we need to know about the circuit is the overall action of the i.c., since the D-type flip-flop is too complex to be worth building up from separate (discrete) components.

The i.c. has two main inputs, D and Clock, and outputs Q and $\overline{Q}$. These outputs are always inverse to each other, they can never be identical unless the circuit has failed. The D (for Data) input consists of a logic signal, 1 or 0, and the clock input is the usual clock pulse described in Chapter 3. The usual action of the D-type flip-flop is that the logic signal voltage at the D input is switched to the Q output by the *leading edge* of the clock pulse. For example, if the D input were at logic 1 before the clock pulse arrived, and the Q output were at logic 0, then the output would be switched over to logic 1 at the leading edge of the clock pulse. Most designs of D-type flip-flops 'lock-out' data during the high part of the clock pulse. This means that the switchover is completed by the time the clock pulse has reached logic 1, and any change of voltage at the D input has no effect on the Q output during the rest of the cycle, the high part of the clock pulse, the trailing edge and the time between pulses.

The D-type flip-flop is very useful as a latch because rapidly-changing data signals can be 'captured' at the leading edge of each clock pulse to the D-type. A digital figure read-out, for example, must not change rapidly while it is being viewed, so a latch of some sort may be needed to keep the displayed figures steady while a reading is taken. The latching action enables a bit of data to be stored for a time equal to the time between clock pulses, no matter how the input signal varies in this time.

A very important circuit action can be achieved by connecting the $\overline{Q}$ output back to the D input as shown in *Figure 4.2* and using the clock input only. Remember that the $\overline{Q}$ output

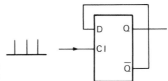

Figure 4.2. Toggling connection to make the D-type into a binary counter

signal will always be the inverse of the Q output signal, and imagine that the Q output is at logic 0, so that the $\overline{Q}$ output is at logic 1. This sets the D input at logic 1 so that the leading edge of the next clock pulse will switch the flip-flop over, making the Q output change to logic 1. This does not happen instantly, however, there is a delay of around 40 ns, a very short time but long enough to ensure that the clock signal has fully reached logic 1 *before* the outputs change voltage. By the time the Q output has changed to 1 and the $\overline{Q}$ to 0, the clock pulse has reached logic 1, and data is locked out. The logic 0 signal at $\overline{Q}$ therefore has *no effect* on the D input until the leading edge of the next clock pulse. At this next clock pulse, the 0 at the D input will cause the Q output to switch to 0, and the $\overline{Q}$ to 1, but again with enough delay to ensure that the change at $\overline{Q}$ has no effect on the D input.

Clock	Q_{n-1}	Q_n	D_n	$\overline{Q}_n$
1	0	1	0	0
2	1	0	1	1
3	0	1	0	0

Figure 4.3. Truth table for a toggled D-type. The Q output changes over at each clock pulse

Clock input and Q output signals for a D-type connected in this way are shown in *Figure 4.3*. Two complete pulses at the clock input produce one complete pulse at the Q output, so that this circuit is variously known as a *scale-of-two, divide-by-two* or *binary* counter.

The D-type flip-flop becomes more useful if two additional inputs are available. One of these is labelled 'set' or 'preset' and is used to force the Q output to logic 1 no matter what the state of the clock pulse happens to be. The other input is labelled 'reset' or 'clear', and its action is to force the Q output to logic 0. These inputs are sometimes called *asynchronous inputs*, because a signal to either of these inputs need not be

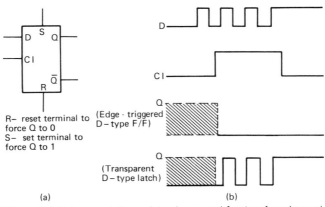

R— reset terminal to force Q to 0
S— set terminal to force Q to 1

(a)

(b)

Figure 4.4. D-type variations: (a) using reset (clear) and set (preset) terminals to fix the Q outputs irrespective of the clock, (b) the D-type edge-triggered flip-flop compared to the transparent latch. The output of the latch will correspond to the state of the D input for as long as the clock is at logic 1. The shading indicates where the Q output may be 1 or 0 depending on how the output was previously set

synchronised in any way with the clock signal. Precautions have to be taken (by gating, for example) to ensure that both inputs are not operated at the same time. Most TTL flip-flops need a negative-going pulse to operate the preset/set and reset/clear inputs.

The name D-type is also used for a type of latch in which binary digits at the D input are gated to the output only while the clock pulse is at logic 1, and are stored (latched) while the clock pulse is at logic 0. This D-type latch cannot be used in the same way as the edge-triggered flip-flop. It would be feasible to construct a D-type flip-flop which was triggered by the trailing edge of the clock pulse, but the leading-edge type is the one generally used.

The J—K flip-flop

The D-type flip-flop, though useful, has several disadvantages which have led to a more versatile circuit, the J—K master-slave flip-flop. The connections to a typical J—K are shown in *Figure 4.5*. Besides the usual clock input, there are two inputs

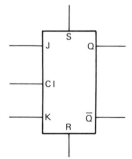

Figure 4.5. J—K flip-flop inputs and outputs. The S (set) input is sometimes labelled P (preset) and the R (reset) input is sometimes labelled C (clear)

labelled 'J' and 'K' respectively which control the action of the flip-flop when the clock pulse arrives. Two other inputs, as noted above, labelled 'set' (or 'preset') and 'reset' (or 'clear') control the output of the flip-flop whether a clock pulse has arrived or not. The usual outputs are Q and $\overline{Q}$. The internal action of a J−K master-slave flip-flop is rather complicated but basically the circuit consists of two flip-flops. The first flip-flop (master) is operated by the *leading* edge of the clock pulse, and its output is controlled by the settings of the J and K inputs at this instant. The outputs of this master flip-flop are connected to the inputs of the second flip-flop (slave), which switches over on the *trailing* edge of the clock pulse. The outputs of the slave flip-flop are Q and $\overline{Q}$, so that these outputs change only at the trailing edge of the clock.

This provides a 'lockout' system which neither depends on the delays in the circuit nor needs such a fast-rising pulse as the D-type. Data at the J and K inputs is transferred to the master at the clock leading edge, and after this time changes at these inputs have no effect. The outputs change at the trailing edge of the clock, so that any connection back from output to input has no effect until the next clock pulse arrives. *Figure 4.6* shows a timing diagram.

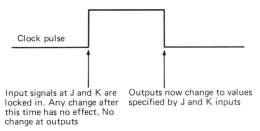

Clock pulse

Input signals at J and K are locked in. Any change after this time has no effect. No change at outputs

Outputs now change to values specified by J and K inputs

Figure 4.6. The timing of events in a J−K master-slave flip-flop

The most significant advantage of the J−K flip-flop is its versatility because of the control exerted by the J and K inputs. *Figure 4.7* shows a truth table, with Q referring to the Q output voltage *before* the clock pulse arrives, and Q_{n+1} the output voltage *after* the clock pulse. Though the truth table summarises what happens, the action is so important that we shall go over it in detail. With J = 0, K = 0, the logic signal at the Q input is unchanged at the clock pulse, so that the flip-flop can be set to latch for as long as J = 0, K = 0.

J	K	Q	Q_{n+1}
0	0	0	0
		1	1
0	1	0	0
		1	0
1	0	0	1
		1	1
1	1	0	1
		1	0

Figure 4.7. Truth table for a J—K master-slave flip-flop

With the inputs set at $J = 1$, $K = 0$, the output Q will switch to logic 1 when clocked (or remain at 1 if that was its previous value). Input settings of $J = 0$, $K = 1$ will cause the Q output to switch to logic 0 when clocked or to remain at 0 if that was its previous value.

When the J and K inputs are set to $J = 1$, $K = 1$, the Q output voltage will reverse at each clock pulse, producing the same counting action as the back-connected D-type flip-flop of *Figure 4.2*, so that the J—K flip-flop can be used as a binary counter without the need to make any feedback connections.

As noted earlier, the asynchronous inputs labelled 'set/preset' and 'reset/clear' can be used to switch the Q output to either logic voltage irrespective of the clock voltage.

Three-state latches

A variation of the latch is the *three-state latch*, a circuit which is particularly important in microprocessor circuits. A three-state latch performs the usual latching function of retaining a signal at its output after the inputs have changed, but with the addition of an *enable* pin. When the enable pin voltage is switched, the outputs of the latch are isolated, floating, and free to take up either logic voltage. This 'third state' is needed when signals can pass along lines in either direction, and avoids the difficulties which could arise if one device were at logic 1 and another connected to the same line were at logic 0. The enabling signal must be generated so that the latch is 'floated' whenever the lines are being used for other signals. Quad (four lines), hex (six lines) and octal (eight lines) latches are available in three-state form, and 'enable high' or 'enable low' types can be selected. Examples include the 74125 and 74126 (quad); 74365 and 74367 (hex).

Binary counters

A *binary counter* is a circuit whose input is a series of pulses and whose output is binary digits, with a separate line for each power of two, 2^0, 2^1, 2^2, 2^3 and so on. The simplest type of binary counter is a chain of connected flip-flops such as that

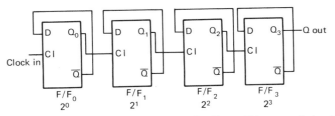

Figure 4.8. A binary counter chain of flip-flops. This type of circuit is known as a ripple (or asynchronous) counter

shown in *Figure 4.8* (in which D-type flip-flops are shown). A counter of this type is variously known as a *ripple, ripple-through* or *asynchronous* counter.

Imagine that we have a chain of flip-flops which change state on the *trailing* edge of the clock. At each flip-flop Q output an l.e.d. is connected so that the l.e.d. will glow when Q = 1. Imagine that all the l.e.d.s are extinguished at the start of counting. When the first pulse arrives, flip-flop 0 is switched over, so its Q_0 output is high (*Figure 4.9*) and its l.e.d. glows. Since this type of flip-flop switches over on the *trailing* edge of the clock pulse, flip-flop 1 is not switched over by the rise of Q_0 from 0 to 1. A second pulse into flip-flop 0, however, will cause this flip-flop to switch again, so that its output voltage

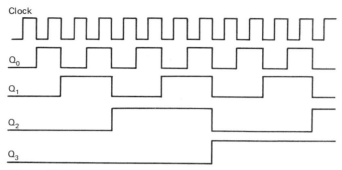

*Figure 4.9. Timing diagram for a binary counter, assuming that all
the flip-flops were reset (cleared) before starting*

drops from 1 to 0. This is a trailing edge, so flip-flop 1 is
switched and its Q_1 output rises to logic 1, lighting the second
l.e.d. At the same time, the first l.e.d. is extinguished by the
drop in voltage of Q_0. A third pulse in will now cause the first
l.e.d. to light again, leaving the second l.e.d. unaffected. A
fourth pulse will extinguish l.e.d.s 1 and 2 but switch on l.e.d. 3.

The reason for numbering the flip-flops 0, 1, 2 rather than 1,
2, 3 now becomes apparent. Flip-flop 0 counts the binary
columns 2^0, flip-flop 1 counts 2^1 etc., so that the l.e.d.s can
be read as a set of binary 1s, each showing a different power of
2 as indicated in *Figure 4.10*. If a gate is now added to the
system (*Figure 4.10*) the counter is nearly complete. With all
flip-flops set to give 0 output (no l.e.d.s lit) the gate is opened
(enabled). Later the gate is closed and the number of pulses

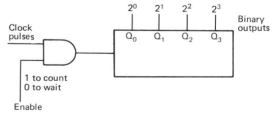

*Figure 4.10. Adding a gate to the binary counter so that
counting can be stopped and started by an 'enable' voltage*

that have passed through the gate will be recorded by the l.e.d.s
of the counter. We can also regard this as a memory circuit
which can be set to a binary digit (read from the most significant digit downwards) which will remain until either it is
reset or more pulses are added. Another way of looking at the
circuit is as a serial *adder*, adding the number of pulses at the
input to any number that has been set by a previous input.

The circuit may be used purely as a binary counter, with
any number of flip-flops — a common number is four or eight.
Alternatively, the circuit may be modified to count in other
ways. Imagine, for example, that a BCD counter is needed.
Such a counter must produce a binary count for up to nine
pulses, but must reset to 0 at the tenth pulse, passing on this

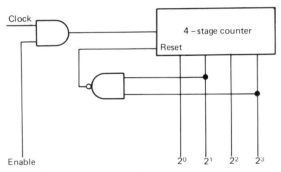

*Figure 4.11. Converting a four-stage counter into a BCD
counter by resetting at the instant when decimal 10
(binary 1010) is detected at the outputs*

tenth pulse to the next BCD stage as a 'carry'. For a count
up to nine, four flip-flops will be needed, and if J-K-type
master-slave flip-flops, such as the 7476, are used the resetting
can be carried out by using the reset (clear) inputs. How can
we use the tenth pulse to operate the reset? The simplest
solution is to use a gate circuit which will detect decimal 10 and
give an output which operates the reset. The reset action of
the flip-flop requires a negative-going pulse on the reset pin,
so that a gate which will have a 1-to-0 change when decimal
10 is counted can carry out the resetting action that we need.

Such a circuit is shown in *Figure 4.11*. Since decimal 10 is binary 1010, a NAND gate operated by the 2^3 and 2^1 outputs of the counter will produce a 1-to-0 change which can be used to operate the reset. Whenever the reset operates, the voltage on the reset pin must then be restored so that counting can restart, so the reset pin is pulsed only very briefly negative. An inverter can convert this brief negative pulse into a positive pulse which can be counted by the next set of BCD flip-flops.

Using gates to control the reset inputs of a chain of flip-flops is a simple and straightforward way of adjusting the maximum count figure, but it is not an ideal method. When high counting speeds, or a large number of stages, are to be used then the time delay for a pulse to 'ripple through' becomes excessive. For example, consider a counter with eight flip-flops, when the count has reached 01111111. The next pulse in will change the least significant digit from 1 to 0, and this in turn will change the next digit from 1 to 0 and so on, until the last digit is changed from 0 to 1. Now there is a measurable delay, 25 ns or more, between the action of the clock pulse and the change at the output, so these changes are not immediate but follow on each other, hence the expression 'ripple-through'.

It would be quite possible to have a circuit arranged to detect the state 10000001 by connecting the 2^7 and 2^0 outputs to a NAND gate input which in turn activated reset. If the clocking rate were fast, however, the early stages of the circuit could have counted several more pulses before the final flip-flop switched over, so that the actual count would not be consistent. The circuit would behave well at low pulse rates, but would give erratic counting at high speeds due to the delays. For this reason ripple counters are not used for applications of this type, though they have one outstanding advantage for high-speed counting. Their advantage is that only the first flip-flop servicing the least significant digit has to run at the high speed of the input pulses. The next flip-flop (2^1) runs at half input speed, and so on. If the counter is used purely as a precise frequency divider, or to register the number of counts on a binary readout *after* pulses have been shut-off, then the ripple counter is ideally suited.

A ripple counter need not necessarily count up, that is from 0 up to whatever maximum value is set by the design. A down-counting ripple counter can be constructed simply by connecting the $\overline{Q}$ outputs of each flip-flop to the clock input of the next in line, as shown in *Figure 4.12*.

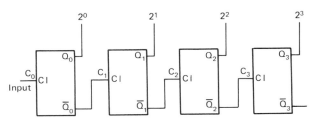

Figure 4.12. A down-counter constructed by connecting each $\overline{Q}$ output to the next clock input. Only the $\overline{Q}$-to-clock connections are shown here

Imagine that the Q outputs have started at the number 1111 (binary), which is 15 (decimal). The first complete pulse into C_0 will switch Q_0 from 1 to 0 and $\overline{Q}_0$ from 0 to 1. The reading of the outputs is now 1110 (remember that C_0 carried the least significant bit) which is 14 (decimal). The next complete pulse will switch Q_0 high again, but with $\overline{Q}_0$ changing from 1 to 0, C_1 will be triggered so that its output falls from 1 to 0. The read-out is now 1101, decimal 13. This countdown will continue until the outputs reach 0, when the next pulse resets the counter to 1111.

A down-count is often a much more convenient method of obtaining an output after a stated number of counts. *Figure 4.13* shows a series of flip-flops arranged as a down counter but with switches connected to the set/reset terminals. These switches permit each counter to be preset to either 1 or 0, so that a binary number can be set up (or *loaded*) on the counter. All that is needed now is a method of detecting the *end* of the count. One method is to use a gate to detect when all the flip-flop outputs are 0. Another method is to use a diode to detect the 0-to-1 change which takes place on the most significant bit when the counter resets to the all-1 stage just after the

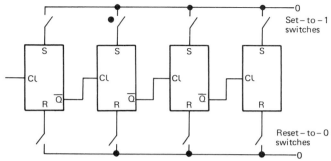

Figure 4.13. A pre-settable down-counter. Each flip-flop can have its Q output set to 0 or 1, so that the down-count starts from this set number

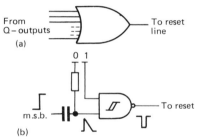

Figure 4.14. Ways of detecting the all-0 state: (a) using a multi-output OR gate whose output is 0 only when all inputs are zero, (b) using a differentiating circuit to detect the positive pulse at the end of a count

all-0 state. This pulse can be sharpened by a Schmitt gate and used to stop counting and to activate other circuits.

Down counting is a much more flexible system because the state which is to be detected is always the same — the all-0s or the change to all-1s. The number of the count is decided by the binary number which is set up on the counter before starting. This is, in fact, a primitive type of programming in which the count is decided by the number loaded (*software*) rather than by the connections within the counter (*hardware*).

The next step that is possible with a ripple counter is to enable counting to go in either direction. Since the only difference between the up-counts and the down-counts is the use of Q for up-counts and $\overline{Q}$ for down counts, gates can be connected between counter stages to route the selected output to the next clock input. *Figure 4.15* shows one possible

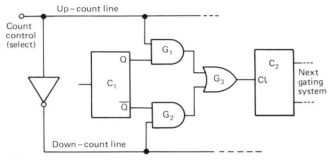

Figure 4.15. An up/down counting system, showing only one pair of flip-flops. The two control lines are driven by a single input using an inverter to ensure that one line is always the complement of the other

system. Each output of one flip-flop is taken to the input of a two-input AND gate whose other input is connected to a line. The AND gates which have inputs from the flip-flop Q terminals are connected to the up line, and the AND gates which have inputs from the flip-flop $\overline{Q}$ terminals are connected to the down line. The AND gate outputs are connected to an OR gate which in turn drives the clock input of the next flip-flop.

When the up line is set (by a control switch) to logic 1 with the down line at 0, then a 1 at the Q output produces a 1 at the output of the AND gate G_1 and, in turn, a 1 at the output of its OR gate, G_3. In this way clock pulses are connected from the Q output of C_1 to the clock input of C_2. If the up line is switched to logic 0, then the output of the AND gate G_1 remains at 0. With the up line held at logic 0 and the down line at logic 1, a 1 pulse at $\overline{Q}$ will cause a 1 output from G_2

which also will cause a 1 output from the OR gate. In this condition the clock pulses into C_2 have been routed from the $\overline{Q}$ output of C_1, so that the count is downwards.

Because the up and down lines must always be inverse, i.e. can never be allowed to have the same logic voltage, a single control can be used with an inverter to ensure the correct line voltages, as shown in *Figure 4.15*. Using this scheme, a 1 on the 'select' input will ensure down counting and a 0 on the 'select' input will ensure up counting.

A complete i.c. up/down binary four-stage counter using such a scheme is the 74191. The up/down input pin is set to logic 0 for the up count and to logic 1 for a down count. Provision is made for each stage of the counter to be loaded,

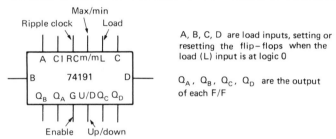

A, B, C, D are load inputs, setting or resetting the flip-flops when the load (L) input is at logic 0

Q_A, Q_B, Q_C, Q_D are the output of each F/F

Figure 4.16. The 74191 counter which permits presetting of each stage, up or down counting, and has various outputs which mark the progress of the count

using the data inputs, A, B, C, D. These are loaded into the flip-flop only when the load pin is taken to logic 0; the data inputs are ignored at any other time. An additional control, *enable*, permits the count to be interrupted. Counting takes place when the voltage on the enable pin is at 0 and is suspended when the enable pin voltage is at 1. The use of enable does *not* reset the counter; if enable is taken high during a count, the count will continue whenever the enable input is taken low again.

Two useful outputs from this particular chip are the *max/min* and *ripple clock* pins. The max/min output is a positive-going pulse longer than the clock pulse at each change-over of the most significant bit of the count − the max/min pulses are arranged so that one has a trailing edge coinciding with the leading edge and the next has its trailing edge coinciding with the trailing edge of the Q_D pulse (*Figure 4.17*). This can be

Figure 4.17. The max/min output of the 74191

used for detecting the end of a count, maximum or minimum. The ripple clock output is a negative-going pulse at the end of a count (but of normal clock pulse width) which can be used as a carry to the next stage of counting.

Synchronous counters

The disadvantages of the ripple counter have been noted earlier, in particular the problem of the time taken for a pulse to ripple through all the stages of a counter. An alternative method of counting is the *synchronous counter* in which each flip-flop is clocked by the same input pulse so that all changes occur at the same time. The design of the synchronous counter with a few stages is fairly straightforward, but the procedure becomes very difficult when a large number of stages is needed. Fortunately, LSI synchronous counters are available for the most-needed number counts, and there are alternative methods available for count numbers which are not decimal, BCD, nor convenient powers of two.

The principle of the synchronous counter is that at each state of the count the binary numbers present on the outputs of each flip-flop at each count number are used to set the J and K inputs of the next flip-flop. The pattern of connections must be correct if the count is to proceed normally.

To see how this operates, consider the very simple two-stage counter of *Figure 4.18*, and recall the truth table for the $J-K$ flip-flop (*Figure 4.7*). The state-table alongside the diagram of *Figure 4.18* shows the Q_0 and Q_1 outputs, and also the voltages present of J_1 and K_1.

The first clock pulse will have the effect of switching over Cl_0, since J_0 and K_0 are permanently connected to logic 1.

State diagram

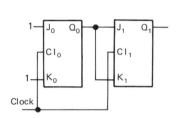

Cl	Q_0	Q_1	J_1	K_1
0	0	0	0	0
1	1	0	1	1
2	0	1	0	0
3	1	1	1	1
4	0	0	0	0

Figure 4.18. A two-stage synchronous counter and its state table. The clock pulses are fed to each flip-flop, and the J K connections determine the action at each clock pulse. In the state table the values of Q_0 and Q_1 are those after the trailing edge of each clock pulse

Because Q_1 (at logic 0) connects to J_1 and K_1, however, Cl_1 will be unaffected by the clock pulse. Q_0 does not switch to logic 1 until the trailing edge of the clock pulse arrives, by which time J_1 and K_1 have 'locked-out' — they were at logic 0 at the *leading* edge of the clock pulse so that this is the setting which decides the output from Cl_1.

At the next clock pulse, of course, J_1 and K_1 are at logic 1, so that the clock input which causes Q_0 to change from 1 to 0 also causes Q_1 to change from 0 to 1, giving the binary count 10 (decimal 2). This leaves J_1 and K_1 set to logic 0, so that the next clock pulse does not affect Cl_1, but changes Cl_0 from 0 to 1. The binary count is now 11, decimal 3. With J_1 and K_1 set to 1 by Q_0, the next clock pulse will switch both flip-flops back to 0 output.

To extend the synchronous counter to more than two stages gating must be used in addition to the flip-flops. A three-stage

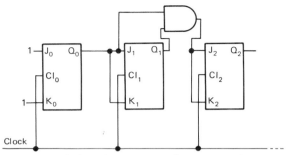

Figure 4.19. A three-stage synchronous counter

counter is shown in *Figure 4.19* with an AND gate used to set the J and K inputs of the third stage. The effect of this gate is that J_2 and K_2 are set to logic 1 only when Q_0 *and* Q_1 are both at logic 1.

The count therefore proceeds in the way described for the two-stage counter until $Q_0 = 1$ and $Q_1 = 1$. At this point, the action of the AND gate holds J_2 and K_2 at logic 1, so that the next clock pulse, which sets Q_0 and Q_1 back to 0, will switch over Q_2, giving the count number 100 (decimal 4). The next three pulses have no effect, because the output of the AND gate is low, so that Q_2 cannot switch over. At decimal 7, however, Q_0 and Q_1 are again both at logic 1, so that the eighth pulse once again switches over Cl_2.

Figure 4.20 shows a four-stage synchronous counter; the reader is invited to check out the gating to the fourth stage

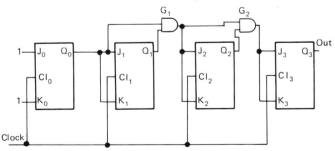

Figure 4.20. A four-stage synchronous counter

for himself. Synchronous counters, like ripple counters, can be set to count in either direction, and a down-count with detection of 0 is, as before, a simpler method of stopping a count after a definite number. Synchronous counters can be designed for resetting after a given number and this can be achieved by gating the J and K inputs only, but the procedure is far from simple when the counters contain large numbers of stages. The usual design method is to write the state of each Q output for each count number, then the J and K numbers for the *next* state, so that gates can be arranged to produce the right combination of J and K settings. It's a tedious process, and the graphical version, called *Karnaugh mapping*, is only really helpful when fairly small numbers of stages are being designed, or when codes other than straightforward binary are to be used.

Counters seldom have to be constructed from flip-flops, since ready-made i.c. counters are available. In addition, shift registers (see Chapter 5) can be used as counters for most counting scales. Nevertheless, constructing a counter from flip-flops is a useful exercise because the action can be checked and traced — this is seldom possible when an i.c. is used!

Practical work

Practical work on counters requires a source of clock pulses. A debounced push-button switch is useful when the action of each clock pulse needs to be examined; a low frequency oscillator is more useful when higher speed clocking is needed. One of the most useful practical introductions to counter techniques is the 7476 dual J–K flip-flop — this i.c. is low-priced, and easy to use. The pinout diagram is shown in *Figure 4.21*. For most low-speed counter applications, the preset and reset pins can be left floating — if they are connected to common lines care must be taken to ensure that they can never *both* be taken low.

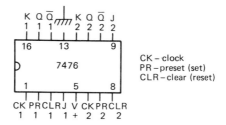

Figure 4.21. Pinout of the 7476 J—K dual flip-flop

The whole range of counters can be built up using 7476 J—K flip-flops; if a four-bit synchronous counter is to be investigated, the 74168 is a synchronous up/down counter. The pinout is shown in *Figure 4.22*. The counting direction is

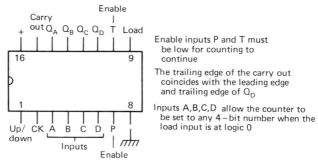

Figure 4.22. Pinout of the 74168 counter

upwards when the up/down pin is at logic 1, and in the downward direction when low. The enable inputs, labelled 'P' and 'T', form a gate which inhibits counting when *both* P and T are high.

5 Shift registers

A *shift register* is a circuit which uses flip-flops connected in line so that each bit which is stored at the Q output is shifted to the next flip-flop in line at each clock pulse. The clock pulses are taken to all of the flip-flops in the register, so that the action is synchronous. Any type of flip-flop whose action is suitable (edge-triggered rather than the latching or 'transparent' type) can be used, and *Figure 5.1* shows typical circuits making use of edge-triggered D-types, clocked R–S types, and master-slave J–K flip-flops.

To understand the operation of these circuits, recall the truth table for each flip-flop, together with the 'lockout' action which ensures that a switchover of the output occurs too late after the leading edge of the clock pulse to affect the next flip-flop, whatever type of flip-flop from the list above is used. Imagine that all the circuits of *Figure 5.1* have been reset so that each Q output is set to 0, and that the data input of flip-flop 0 is at logic 1 in each case.

The D-type flip-flop, with a 1 at its D input, will switch over at the leading edge of the clock pulse, but the delay will ensure that its Q output does not affect the input of the next flip-flop and so alter the Q_1 output after the leading edge of the clock pulse. Similarly, the R = 1, S = 0 (because of the inverter) input of the R–S type ensures that $Q_0 = 0$, $\overline{Q}_0 = 1$ shortly after the leading edge of the clock pulse, but with no effect on flip-flop 1. The J–K is similarly affected, with $J_0 = 1$, $K_0 = 0$ at the time of the first clock pulse, $Q_0 = 1$, and $Q_1 = 0$ at the trailing edge of the clock, setting up the second J–K flip-flop for the next clock pulse.

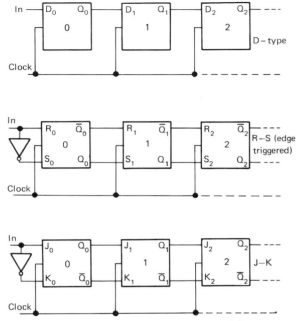

Figure 5.1. Shift-register circuits, using three different types of flip-flop. Note that it is the $\overline{Q}$ output of the R−S flip-flop which drives the next R input

Whatever type of flip-flop circuit is used, therefore, the action of the first clock pulse on the first flip-flop has been to switch the output from the original setting of logic 0 to logic 1, the voltage of the data input. At this first clock pulse, each of the other flip-flops in the chain has had a data input equal to zero from the previous stage (R = 1, S = 0 for the R−S type) so that the first clock pulse leaves each flip-flop except the first unchanged with zero output. See *Figure 5.2*.

At the second clock pulse, however, with the data input to the first flip-flop still at logic 1, the data input to each of the second flip-flops is also 1 (R = 0, S = 1 for the R−S type). At the second clock pulse, therefore, the output of flip-flop 1 will be unchanged, but the output of flip-flop 2 will switch over to logic 1. The other flip-flops which follow will be unaffected because at the leading edge of the clock pulse they have both had zero input.

At the third clock pulse, assuming that the input is still held at logic 1, the third flip-flop of each chain will switch over, leaving the fourth at logic 0. The fourth flip-flop will, in turn, switch over on the fourth clock pulse.

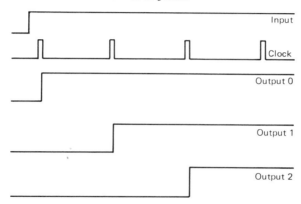

Figure 5.2. The shift action of the register illustrated

Shift registers form a very important class of components in all types of digital circuits. Because the flip-flop output is changed only by a clock pulse after the input has been changed, the removal of clock pulses, leaving only the supply voltage present, leaves the output of the flip-flop unchanged for as long as these conditions are maintained. Each flip-flop can therefore be made to *store* a binary digit (or bit) for as long as the power is applied and the clock pulses are gated out. A set of binary digits can be stored in a register which has one flip-flop for each digit. We can expect, then, to find shift registers extensively used wherever there is any reason to store bits from one operation to another, and several typical register applications are detailed in this chapter and also in Chapter 7, which deals with microprocessors.

Before we look at a selection of shift-register circuits, we need to note that there are four basic types of register. The type shown in *Figure 5.1* is a *Serial-In-Serial-Out* (SISO) type. The logic voltage at the input is fed into the shift register at each clock pulse, and can change in the time between clock pulses. After a number of clock pulses equal to the number of flip-flops in the register, the same bit is available at the output. A SISO register used in this way can act as a time delay – the bits at the output are delayed by several clock pulses (equal to the number of flip-flops) compared to the bits at the input.

Another basic type of shift register is the *Parallel-In-Parallel-Out* (PIPO) type (*Figure 5.3*). In this type, the information

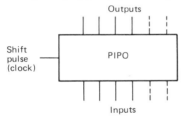

Figure 5.3. A PIPO register

(data) bits are loaded into each flip-flop in parallel, using the set/reset terminals rather than by clocking. If no clock pulses are applied, of course, the bits are not shifted, and the readout at the Q terminals is the same as was loaded in. This use of the register is a convenient method of temporary storage for a few bits. If clock pulses are applied, then each bit will be shifted one place at each clock pulse, and the readout will not be identical to the original data pattern which was loaded in. This leaves the unanswered question of how the serial input and output are used. One possibility is to connect the input to logic 0 or logic 1, and to ignore the output. In this form of operation, the register will fill up with 0s and 1s after a complete set of clock pulses. Another possibility is to rotate, to connect the final output of the register back to the serial input, so that the register output is fed back in at each clock pulse, and a complete set of clock pulses leaves the contents of the register unchanged.

The two remaining types of shift register combine serial and parallel methods. The *Parallel-In-Serial-Out* (PISO) register (*Figure 5.4*) uses the set/reset terminals of a J—K flip-flop to load data bits into each flip-flop independently and at the same time. The data is then shifted out one bit at a time when

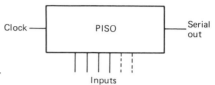

Figure 5.4. A PISO register

clock pulses are applied. This makes it possible for data which was present in parallel form (on several lines at the same time) to be converted to serial form (one bit after another) to be transmitted along a single line. Conversion to serial form is necessary if the bits are to be transmitted along telephone lines, recorded on tape, sent to a video monitor (see Chapter 6) or used to operate a teleprinter.

The *Serial-In-Parallel-Out* (SIPO) register (*Figure 5.5*) performs the opposite function. In this type data is presented a bit at a time and is shifted in at each clock pulse. After a complete set of clock pulses the register is full, and the contents can be read at the Q terminals or unloaded along a set of parallel lines. In this sense 'unloaded' means simply that the

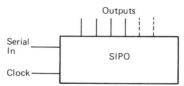

Figure 5.5. A SIPO register

bits can be used to operate gates or other circuits — the registers are not changed in any way by this action. Using the SIPO register, data bits which have been transmitted in sequence from a single line can be collected to make a 'word' of several bits.

One simple example of the use of a SIPO register is as a counter with decimal indication, using a 10-stage shift register.

Logic 1 is loaded into the register on the last flip-flop, and logic 0 into all of the remaining flip-flops. The serial output is then coupled back to the serial input, as shown in *Figure 5.6*, and each parallel output is connected to a display digit (see Chapter

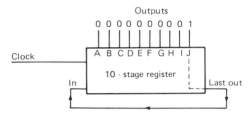

Figure 5.6. A decade (decimal) counter using a 10-stage register − this is an example of a ring counter

6 for details). As shown, before a clock pulse arrives the 1 stored in the last flip-flop illuminates the 0 of the display. The first clock pulse will then shift the logic 1 into flip-flop A so that the output of F/F A is now logic 1, illuminating the figure 1. Similarly, the second clock pulse will shift the 1-bit so that digit 2 is displayed, and so on. A carry to a second stage of counting can be arranged by connecting the output of the zero counter to the clock input of another similar register. Since a 10-stage shift register can be made as a single-chip i.c., this method of decoding a decimal count is simpler than the binary-to-BCD methods which will be described later, but only if a suitable display system is available.

Johnson counters

The *Johnson counter* is a circuit which uses shift registers to provide synchronous counting without design complications. High counting speeds are possible, and the decoding can be comparatively simple, unlike a binary or BCD counter. The principle of the Johnson counter is to connect a shift register with the final output connected back to the serial input, but with inversion (*Figure 5.7*). Because of the appearance of the circuit the counter is sometimes referred to as a *twisted-ring counter* to distinguish it from the *simple ring counter* of *Figure 5.6*. The simple Johnson counter has a maximum count length of 2n, where n is the number of flip-flops used in the shift

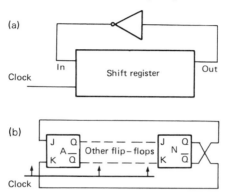

Figure 5.7. The Johnson counter (a) is also known as the 'twisted-ring' counter because of the inversion (b) which can be carried out by crossing over the Q-J and $\bar{Q}$-K connections between the last flip-flop and the first. When an i.c. shift register is used, this inversion has to be performed by a separate inverter

register. This compares with 2^n for a binary counter, and n for a simple ring counter.

The action of a Johnson counter, as illustrated in *Figure 5.8*, is as follows: Imagine that the starting condition is with all the outputs at zero. The inversion of the final output will therefore place a logic 1 at the input, so that the first clock pulse loads a 1 into F/F A. At the second clock pulse the input conditions of F/F A are unchanged and the 1 in F/F A will be shifted into F/F B. This action continues, loading 1s into F/F A each time, until five clock pulses (in this example) result in the register being filled with 1s. The inversion of the final output now ensures that a logic 0 is placed at the serial input, so that the next clock pulse will load a 0 into F/F A and each successive clock pulse will shift the 0 across and load in another 0 until the register is once again filled with 0s. *Figure 5.8* shows the truth table, which also shows the outputs to be gated in order to detect each stage of the output. Note that each count number can be detected by a two-input gate.

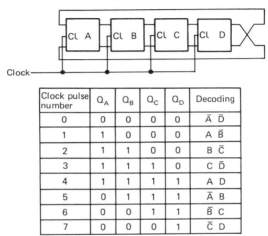

Clock pulse number	Q_A	Q_B	Q_C	Q_D	Decoding
0	0	0	0	0	$\bar{A}\ \bar{D}$
1	1	0	0	0	$A\ \bar{B}$
2	1	1	0	0	$B\ \bar{C}$
3	1	1	1	0	$C\ \bar{D}$
4	1	1	1	1	$A\ D$
5	0	1	1	1	$\bar{A}\ B$
6	0	0	1	1	$\bar{B}\ C$
7	0	0	0	1	$\bar{C}\ D$

Figure 5.8. A four-stage Johnson counter and its truth table. The decoding column shows what quantities are AND-ed to form the eight outputs

The simple Johnson counter, because the count number is equal to twice the number of flip-flops used, can be used for even number counts only, since twice any number is always an even number. To count odd numbers, gate circuits must be added so that the all-1 state is skipped in each cycle. This can be done by using the gate to detect the state just before the all-1s state. For a five-stage register this state is 11110, so a four-input AND gate and inverter (to invert the 0) are AND-ed to the input. In the example of nine-counter shown in *Figure 5.9* NAND gates are used, so the input has to be inverted again.

Since the flip-flops of a shift register may go to the logic 1 or to 0 when power is applied, some method of setting a

Johnson counter is needed before counting can start. If the shift register is of a type which allows each stage to be cleared (reset), this is a simple solution, particularly if the reset can be automatically carried out when the power is switched on. Circuits which do this (based on the use of a monostable) are called *initialising circuits*. If this method cannot be used, gating must be arranged so that the counter will settle into its correct pattern. To ensure that a counter of more than three stages is self-starting without a clear action, feedback is needed from more than one stage. For design details, the interested reader is referred to Texas Instruments' *Application Report B102*.

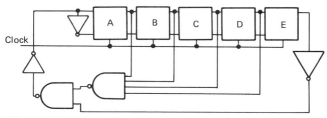

Figure 5.9. A five-stage Johnson counter with gating to detect the state 11110, and so switch a 0 into the input in place of the 1 from stage E which would normally cause the count to go to 11111

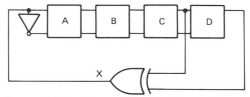

Figure 5.10. A maximum-length non-binary counter. The counter flip-flops would normally be part of an integrated shift register

Other types of synchronous counters using shift registers are possible and usually easier to design than binary synchronous counters, particularly when a long count number is needed. The maximum count length which can be obtained by such methods is only one less than that of a binary counter using the same number of flip-flops.

SHIFT REGISTERS

The circuit shown in *Figure 5.10* uses feedback through an X-OR gate to the input of the register. The flip-flop outputs which have to be used to feed the X-OR gate depend on the number of flip-flops in the register; for three or four stages, only the last two stages (the most significant bits) need to be used. The circuit for a four-stage counter is shown in *Figure 5.10* and the count sequence is shown in *Figure 5.11*. Note

Count	A	B	C	D	X
1	1	1	1	1	0
2	0	1	1	1	0
3	0	0	1	1	0
4	0	0	0	1	1
5	1	0	0	0	0
6	0	1	0	0	0
7	0	0	1	0	1
8	1	0	0	1	1
9	1	1	0	0	0
10	0	1	1	0	1
11	1	0	1	1	0
12	0	1	0	1	1
13	1	0	1	0	1
14	1	1	0	1	1
15	1	1	1	0	1

Figure 5.11. The count sequence for the circuit of Figure 5.10

that the counting action will stop if the register becomes filled with 0s at any time, so that a preset (set) input or a gating circuit must be used to prevent this from happening at switch on.

To understand the action, imagine the register filled with 1s. Since the inputs to the X-OR gate are both 1s the output of the gate is 0, and this 0 will be shifted into the first flip-flop of the register. The next counts have the same action, but at the fourth count the shift action has left only a single 1 input to the X-OR gate. With this input, the output of the X-OR gate is also 1, so that a 1 is loaded in at the next count. Meantime, the inputs to the X-OR gate have gone to 0, so that the next digit loaded in is 0. This cycle then continues until the first repetition after a count of decimal 14.

Figure 5.12 shows what stages are needed to feed the X-OR gate for various numbers of stages of shift registers up to 12. Shorter counts can be obtained by using additional gating which causes the counter to jump one or more states. The gating is chosen so that the jump needs only a different digit loaded into the input. For example, to shorten the count of the circuit of *Figure 5.10* by three places, the count can jump from 1000 to

Number of stages	Stages fed to X-OR gate	Count
3	B C	7
4	C D	15
5	C E	31
6	E F	63
7	F G	127
8	D E F H	255
9	E I	511
10	G J	1023
11	I K	2047
12	F H L	4095

Figure 5.12. Connections from the shift register outputs to the X-OR gate for making counters of various lengths

1100 by loading in an extra 1. The state 1000 is the only state in which the outputs of the flip-flops B, C and D are all zero, so that a gate which detects this state can be OR-ed with the X-OR gate to provide the skip. For longer counts the gating can sometimes be simplified by using Boolean algebra to describe the jump condition.

Binary, decimal and BCD conversion

Because of its simplicity, binary arithmetic is used extensively in digital circuits, but displays are generally of decimal numbers (see also Chapter 6). To display decimal numbers some sort of conversion of binary numbers into BCD is needed. This is not difficulty to carry out, using a shift register, provided that the rules for converting binary into BCD are known. The rules are:

(1) The binary number is shifted, starting with the most significant bit, into the shift register, whose flip-flops are grouped in fours.

(2) Each group of flip-flops in the BCD register represents a power of 10 in a decimal number, with the binary bits shifted into the lowest order group first.

(3) If, before a shift pulse, a group of four flip-flops contains a number less than five (decimal), shifting takes place normally.

(4) If the number stored in a group of four flip-flops is five to nine inclusive, binary three (0011) is added to the stored number before shifting.

Figure 5.13 shows an example of these rules being applied. The number to be converted is binary 11010 (decimal 26) and we can imagine the bits shifted in, most significant bit first, into a pair of four-bit shift registers which will eventually contain the BCD numbers. At the first shift, the 1 bit moved into the lower BCD register produces a total of one in this register, and the second bit shifted in produces a count of three. This does not require any correction, but the next bit shifted in, a 0, makes the total 110 (decimal 6) which now requires the correction in the form of adding 011 (decimal 3). This makes the total in the register equal to 1001. The next shift pulse

	Shift registers		*Number*
	8 4 2 1	8 4 2 1	
			1 1 0 1 0
1st shift	0 0 0 0	0 0 0 1	1 0 1 0
2nd shift	0 0 0 0	0 0 1 1	0 1 0
3rd shift	0 0 0 0	0 1 1 0	1 0
Add 3	0 0 0 0	0 0 1 1	1 0
Total	0 0 0 0	1 0 0 1	1 0
4th shift	0 0 0 1	0 0 1 1	0
5th shift	0 0 1 0	0 1 1 0	

Figure 5.13. Binary-to-BCD conversion, using the 'add-3' method

moves the 1 across to the next BCD register, but with no correction needed. A 1 is also shifted into the lower register, but again no correction is needed. The final stage of shifting moves a 0 into each register, and since this is the last pulse (number of clock pulses equals number of digits in the number to be converted) no addition is needed to the number in the lower register even though it has reached 0110 (decimal 6). The figures in the register are now 0010 0110, the BCD equivalent of 26.

To make these rules operate in the form of hardware, three types of circuit are needed. One is, as might be expected, a shift register, one is an adder (the action of an adder is covered in Chapter 7), and the third is a gate which detects when the content of a group of four BCD digits is five or more.

Figure 5.14 shows the circuit of a four-decade binary-to-BCD converter, courtesy of Texas Instruments. The shift register is a 74199, which is an eight-bit PIPO type. Pin 23 on this particular circuit is a shift/load input; a logic 1 level at this pin will cause the i.c. to shift right at each clock pulse, and with a logic 0 input the register can be parallel-loaded at the inputs A to H inclusive. For binary-to-BCD conversion pin 23 is maintained at logic 0.

The binary number is entered at the input of the first 74199, most significant bit first. This can be done by clocking the digits out of another register, using a shift direction which allows this most significant bit to be entered first. At each clock pulse a new bit is entered, and the remaining bits are shifted (to the right in the diagram of *Figure 5.14*), but *through the adder*, as noted below.

The outputs of each group of four are connected to the 7452 gate i.c. This contains four AND gates whose inputs (A and B) are connected to an OR gate which has also an enable input. The inputs from the shift register are connected so that the output from the 7452 will be 1 for (Q_A AND Q_C) OR (Q_B AND Q_C) OR Q_D equal to 1. In Boolean algebra, this is $Q_A.Q_C + Q_B.Q_C + Q_D$. Those are the conditions for the number stored in the four flip-flops of the shift register to total (decimal) 5 or more, because if $Q_D = 1$, then decimal 8 is stored, and if

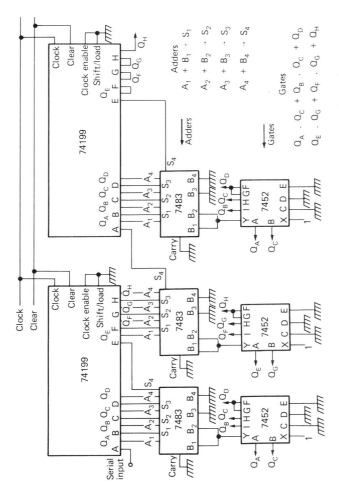

Figure 5.14. Carrying out a binary-to-BCD conversion using shift registers, adders and gates

Q_B and Q_C are each 1, then the number stored is 6. If Q_A and Q_C are each 1, then the number stored is 5. For numbers 7 and 9, Q_C, or Q_B and Q_A, will be 1, so that the condition Q_B and Q_C equal to 1 is fulfilled. If 9 is stored (1001) then $Q_D = 1$ and $Q_A = 1$ so the output of the gate is also 1.

The effect of a 1 output from the 7452 is applied to the B_1 and B_2 inputs of the adder (7483). Now these are the least significant and next higher bits which will be added to the lowest two bits of the number already stored in the shift register, so that the number 0011 (decimal 3) is added to the number stored in the register. The action of the shift/load pin, remember, has been to allow parallel-loading only; the register does not shift at each clock pulse. What is happening is that the digits at the A and B inputs of the adder are added, and the sum presented at the S outputs, with a carry to the next digit along. Shifting takes place because the sum of Q_A and the 7452 gate output is the input to flip-flop B, the sum of Q_B and the gate output is the input to flip-flop C, and so on. When a number of clock pulses equal to the number of bits has been applied, the registers contain the BCD number which is equivalent to the binary number.

Another method of binary-to-BCD conversion uses an i.c. developed for the purpose, the 74185. Since this i.c. is a memory rather than a shift register it will not be dealt with in detail here.

Converting from BCD to binary is equally important, because normally a decimal keyboard will be arranged to give a BCD output of each digit and this output total must be converted into pure binary form unless circuits for performing BCD arithmetic are available. As might be expected from the action which is required to convert binary into BCD, the conversion of BCD into binary involves shifting and subtraction. The digits of the BCD number are, of course, grouped into fours and shifted right at each clock pulse. When a logic 1 crosses from one group of four to the next, decimal 3 (0011) has to be subtracted. This is done by adding the 2's complement form of 3, 1101 in BCD form, and discarding any carry bit. The steps in a typical conversion are shown in detail in *Figure 5.15*. The number to be converted in this example is

0011 0101. *Figure 5.15* shows this written in two groups; in practice, of course, an eight-bit shift register could hold all of the digits, but could be used only if it were possible to perform the addition of 1101 at the appropriate times.

At the first shift, a 1 is moved from the higher digit group to the lower digit group, and another 1 is moved out of the BCD lower register into the answer register. In accordance with the rules of conversion, 1101 is now added to the BCD number in the lower digit group, and the carry is ignored. Note that nothing is added to the digit which has been shifted into the answer register because this is no longer a bit of the BCD number.

	BCD high digit	BCD low digit	*Answer register*
	0 0 1 1	0 1 0 1	
First shift	0 0 0 1	1 0 1 0	1
− 3		1 1 0 1	1
Ignore carry		0 1 1 1	1
2nd shift	0 0 0 0	1 0 1 1	1 1
−3		1 1 0 1	1 1
Ignore carry		1 0 0 0	1 1
3rd shift	0 0 0 0	0 1 0 0	0 1 1
4th shift		0 0 1 0	0 0 1 1
5th shift		0 0 0 1	0 0 0 1 1
6th shift		0 0 0 0	1 0 0 0 1 1
7th shift		0 0 0 0	0 1 0 0 0 1 1
8th shift		0 0 0 0	0 0 1 0 0 0 1 1 final result

Figure 5.15. BCD-to-binary conversion, using the 'subtract-3' method

The next clock pulse shifts another 1 into the lower digit register, and 1101 has to be added to this register again, ignoring the carry. All of the 1s have now been shifted out of the higher register, so that all the remaining steps consist of right-shifts only until all eight pulses have been used. The final result is 00100011, binary 35.

Figure 5.16 shows a circuit which carries out the conversion for a four decade counter. The 74199 is an eight-bit PIPO shift register with shift/load control as described earlier, and the

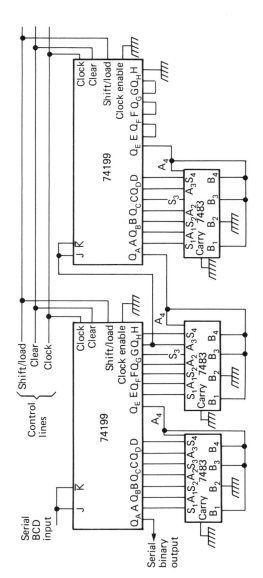

Figure 5.16. Carrying out a BCD-to-binary conversion using shift registers and adders

7483 is a four-bit binary adder. The sequence of operation is as follows:

The registers are cleared, using a 0 on the common clear line, and the shift/load input at pin 23 of each register is taken high, so that there will be a right shift at each clock pulse. The BCD number is then entered serially, most significant bit first, using 16 clock pulses. If the BCD number has been obtained from a keyboard, a pair of eight-bit registers will be needed to hold the numbers until they can be clocked in.

The shift/load input is then set to logic 0, so that the registers are ready for parallel entry. The most significant bit of the four-decade number is now at Q_H of register 2, and the least significant bit is at Q_A of register 1. The first clock pulse will feed out the least significant bit at the binary output (Q_A, register 1) and each bit in the registers will be fed one stage to the left of the diagram because of the connections between outputs and inputs. For example, the most significant bit at Q_H will be fed into the G input, the bit at Q_G into the F input, and so on in the second register. The bit which crosses over into the next group is at Q_E, and this is fed to the adder, using the A_4 input along with the B_1, B_3, B_4 inputs. Now if the bit at Q_E is 0, then the B inputs to the adder are all 0, so that nothing is added. If the bit at Q_E is 1, then $B_4 = B_3 = B_1 = 1$, and B_2 is set to 0, giving the number 1101 to be added to the number at the A inputs. The sum bit of A_4, B_4 is at S_4 and is loaded into the D input, the A_3, B_3 sum at S_3 is loaded into the C input, and so on. There is no carry into the adder, and the carry out is left disconnected so that a carry bit is not used. This adder connection automatically ensures that the correcting factor 1101 is added in each time a 1 bit is taken across a group of four, so that the bit fed out of Q_A, register 1, at each clock pulse is a bit of a 16-bit binary number.

Multiplexers and demultiplexers

Multiplexers and *demultiplexers* are not shift register circuits but are frequently used along with shift registers. A multiplexer, or data selector, enables one bit from a large number of bits at the inputs to be gated to an output by selecting a code on a set of 'select' inputs. A multiplexer with a single output can perform the conversion of parallel-to-serial that is usually carried out by a shift register (SIPO); unlike the shift register, however, the multiplexer is not confined to bits in sequence.

Figure 5.17 shows the pinout of the 74150 16-bit data

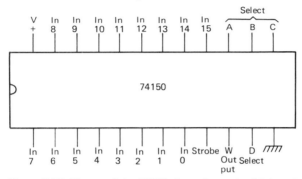

Figure 5.17. Pinout of the 74150 data selector (multiplexer)

selector. There are 16 input pins, each one of which may have a binary bit applied. The strobe signal at pin 0 will either shut off all the gates (strobe logic 1) or permit an output (strobe logic 0), and the four data select pins, 11, 13, 14, 15 are used to 'address' the input which is to be transmitted to the single output pin. A 0 at each data selector pin, for example, gives an address number of 0000, selecting input number 0 (at pin 8) to be connected to the output. Selecting A = 1, C = 0, B = 0, D = 0 gives the address number 0001, and so input 1 is selected. Data select pin D is for the highest order address number, A for the lowest. Several groups of data selectors may be used so that their outputs form a 'word'. For example, four 74150 selectors may be used so that any one of 16 BCD digits may be selected at the outputs, consisting of one bit from each selector, with the address lines paralleled up.

A demultiplexer, or decoder, performs the opposite function. The input lines are a data line(s) and the usual address lines.

For example (*Figure 5.18*), the 74154 decoder has 16 output pins, a data input pin, an enable input and four address inputs. When enable and data inputs are at 0, a 0 will be output from the pin whose number corresponds to the address number on the four address lines. The outputs at all the other pins are at logic 1. As before, the D address pin is the most significant line of the four-bit address, and A the least significant line. A 1 input at the 'data in' pin results in 1 signals from all of the out-

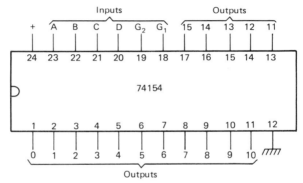

Figure 5.18. Pinout of the 74154 4-to-16-line decoder (demultiplexer)

put pins, and a 1 input at the enable pin also results in an all-1 output, irrespective of the signals at the data pin.

This decoder can also be used as a four-line-to-sixteen-line decoder by connecting the enable and data pins to logic 0. The address inputs can now be used as data input lines so that four input lines can produce up to 10 outputs. The output pin which is addressed will be at logic 0, and all the unaddressed pins at logic 1. Both multiplexing and demultiplexing methods are important in memory selection and addressing, as noted in Chapter 7.

6 Displays and drivers

Counting systems

A digital circuit which is used for any sort of counting will need a display of the number to which the system has counted. The most common forms of display are binary, octal, decimal and hexadecimal. Binary displays consist simply of l.e.d. or similar indicators which show the state of each flip-flop in the counter.

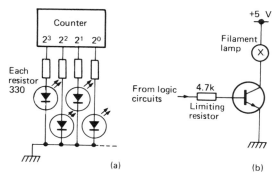

Figure 6.1. Driving binary indicators: (a) direct driving of l.e.d.s, with limiting resistors, (b) driving a filament lamp, using a transistor 'interface'

The circuits which can be used with such indicators are shown in *Figure 6.1*. L.E.D. indicators can be driven directly from TTL counter circuits, using a limiting resistor to prevent over-driving. The voltage across a conducting l.e.d. is about 2.0 V (the precise voltage depends on the semi-conductor material;

different materials are used for different colours of l.e.d.s, and currents of 2 to 10 mA are typical. For larger displays using incandescent bulbs, a transistor interface circuit will be needed, as TTL i.c.s cannot provide enough current for such lamps. A similar arrangement can be used when l.e.d.s are driven by CMOS logic, an alternative is to make the stage which drives the displays from the low-power Shottky TTL series of i.c.s.

Such a binary display is simple, but because of the large number of digits which is needed, and the ease with which 1s and 0s can be transposed when the numbers are written down, simple binary displays are seldom of much practical use.

Octal displays use counting to base 8, in which 7 is the highest figure, and 10 represents (decimal) 8. Converting a decimal number, on paper, to octal is carried out in the same way as that used for binary conversions, except that the successive division is by eight rather than by two (*Figure 6.2*). Octal conversion to decimal is also carried out on paper in the same way as binary-to-decimal, but using powers of eight (*Figure 6.3*). What makes the octal scale attractive is the very simple conversion of binary-to-octal. The binary number is

Remainders

```
8  )275  (
      34  |    3
       4  |    2
     ...  |    4       ↑  Read up
Octal:  423
```

Figure 6.2. Converting decimal to octal. The number is repeatedly divided by eight, with the remainder separated. The remainder forms the octal number, as shown

Power of 8	Decimal
0	1
1	8
2	6 4
3	5 1 2
4	4 0 9 6
5	3 2 7 6 8
6	2 6 2 1 4 4

Figure 6.3. Table of powers of eight

grouped into three-digit units starting at the least significant
bit. Each three-bit number is then written as a decimal number;
since three bits cannot give a number greater than seven, the
converted numbers are in octal scale. Examples of this conversion in both directions are shown in *Figure 6.4*. Another
attractive feature of the octal scale is that the easily-obtainable
BCD-to-seven-segment decoders can be used, with only the
lower three input lines connected (*Figure 6.5*).

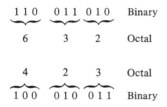

Figure 6.4. Binary-to-octal and octal-to-binary conversions — each octal digit corresponds to three bits

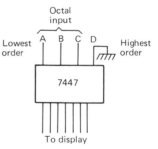

Figure 6.5. Using a BCD-to-seven-segment decoder to display octal numbers

Octal conversion reduces the possibility of errors when many
binary numbers have to be written down, as occurs when a
microprocessor is being programmed. The hexadecimal
representation, using a scale based on 16, gives an even smaller
number of digits but has the disadvantage of needing additional
symbols to represent the numbers 10, 11, 12, 13, 14, 15
(decimal). This is not so inconvenient as it sounds, because
some of the commonly-used display systems (see later, this
chapter) will provide the letters A, B, C, D, E, F (in stylised
form) to represent the extra digits which are needed. In this

Decimal	Hexadecimal
0	0
1	1
2	2
3	3
4	4
5	5
6	6
7	7
8	8
9	9
10	A
11	B
12	C
13	D
14	E
15	F

$X'1F = 16 + 15 = 31$

$X'2C = 2 \times 16 + 12 = 44$

$X'FF = 15 \times 16 + 15 = 255$

1 8 9 7 decimal to hex:

```
16)1897 |
 16)118 | 9
   16)7 | 6
        | 7
```

$= \underline{X'\,769}$

Figure 6.6. The hexadecimal scale, and some conversions between hexadecimal and decimal

scale (*Figure 6.6*) 10 represents (decimal) 16, so to avoid confusion this and other hexadecimal (hex) numbers are written variously as $X'10$, $X10$, or 10_{16}, where the X', X, or subscript 16 indicate that a hex base is in use. Conversion between hex and decimal scales can be achieved (with difficulty!) by the same methods as are used for octal, but using 16 in place of 8. Because our learning of multiplication tables (if we were of the lucky generations) stopped at 10 or 12 times tables, hex is rather more difficult to convert, so that tables such as those in *Figure 6.7* are useful. In practical terms hex is not easy to display; there are very few decoders available and many types of display are difficult to read, mainly because of poor discrimination between B and 6.

Like the octal code, hex is particularly useful because of simple conversion to and from binary. What makes hex even more useful is that a group of four bits makes up one hex character, so that the commonly-used eight bit group, or byte, can be represented by a two-digit hex number. Since the ability of human beings to be confused by digits seems to

Hex number place						Least significant	
4		3		2		1	
Hex	Decimal	Hex	Decimal	Hex	Decimal	Hex	Decimal
0	0	0	0	0	0	0	0
1	4096	1	256	1	16	1	1
2	8192	2	512	2	32	2	2
3	12288	3	768	3	48	3	3
4	16384	4	1024	4	64	4	4
5	20480	5	1280	5	80	5	5
6	24576	6	1536	6	96	6	6
7	28672	7	1792	7	112	7	7
8	32768	8	2048	8	128	8	8
9	36864	9	2304	9	144	9	9
A	40960	A	2560	A	160	A	10
B	45056	B	2816	B	176	B	11
C	49152	C	3072	C	192	C	12
D	53248	D	3328	D	208	D	13
E	57344	E	3584	E	224	E	14
F	61440	F	3840	F	240	F	15

Hex to decimal: Find from table the decimal number which corresponds to hex number for each place, then add.

Example: 0F2C

Place 1	C →	12
Place 2	2 →	32
Place 3	F →	3840
Place 4	0 →	0000
Total		3884

Decimal to hex: Find decimal number, from table, equal to or just less than given number. Note equivalent hex. Subtract decimal difference and repeat.

Example: 2215

Nearest is 2048 = X'800

2215
2048
———
167   X'A0 = 160
160
———
7

Hex number is 8A7
Check: 7 + 160 + 2048 = 2215

Figure 6.7. Hex-to-decimal conversion table for four-digit hex numbers

increase with the number of digits (as the change to all-number telephone systems indicated), working with hex numbers greatly reduces errors, once some practice has been gained.

A few conversions are shown in *Figure 6.8* to show how large binary numbers can be simply converted, and how similar-looking binary numbers convert into hex numbers which are less easy to confuse.

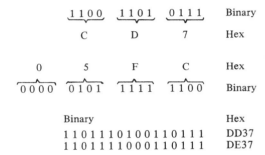

Figure 6.8. Binary-to-hex conversions, with an example of how similar-looking binary numbers are more clearly distinguished in hex notation

For everyday use, however, a decimal display is always needed — no-one is likely to want a digital voltmeter which reads in hex! Apart from a few direct-decimal counter-display systems most decimal displays use 8-4-2-1 BCD as a midway system between the binary arithmetic units and the decimal displays. The types of display which do not use BCD are nowadays seldom used, but will be described briefly.

Display systems

The older types of counting display used systems which displayed a decimal digit when a voltage was applied to a pin, so that one pin was needed for each digit. Incandescent (filament) displays of this type used miniature light-bulb filaments arranged in the shapes of decimal numbers 0 to 9, with one common connection. Selection of a digit is made by applying the working voltage to the other end of the selected filament, so that the filament glows and the glow forms the shape of the digit. Modern filament displays are of the seven-segment type (see later, this chapter) using 5.0 V at around 8 mA for each segment. The modern seven-segment filament display has considerable advantages compared to some of its competitors, notably a wide viewing angle, low current operation, large digit size, low price and long operating life.

Gas-filled display tubes were once a very common method of displaying decimal digits, but today they are seldom used, at least in the smaller sizes. The principle is that if two wire electrodes are surrounded by a gas (such as neon) at low pressure, a suitable voltage (45.0 V or more) between the wires will cause the gas to glow, but the glow surrounds the negative (cathode) wire almost exclusively. By making the cathode wire into the shape of a digit, the glowing gas will also take the form of this digit. In a gas-filled display, 10 sets of cathode wires are made into the form of digits 0 to 9 inclusive. With one common anode wire connected to about +50.0 V, setting a cathode wire to 0 volts will cause one digit to glow. A display of this type needs a set of 10 inputs (11 if a decimal point is included) and each digit is displayed by placing 0 volts on the appropriate cathode. To keep a digit extinguished, however, the cathode voltage of the digit needs to be at a fairly high voltage, typically +30.0 V. As this working voltage is too high for the normal type of TTL or CMOS circuit some sort of interface circuit is needed. This can take the form of a special TTL driver, using 30.0 V rated output stages, or a transistor interface for each cathode line.

Seven-segment and dot-matrix displays

Modern display devices, whatever method of obtaining the display is used, are organised either as seven-segment or as dot-matrix patterns. The seven-segment type, as the name suggests, uses a pattern of seven bars arranged as a slightly oblique figure 8 (*Figure 6.9*). By convention, the letters shown in *Figure 6.9* are assigned to the segments. By lighting up appropriate segments digits can be displayed, and also stylised forms of the letters A to F inclusive. The shapes of the digits are reasonably easy to recognise, but a mixture of upper and lower case lettering must be used to avoid (for example) confusing 'B' with '8', 'D' with '0'. Note the difference (*Figure 6.9*) between 'b' and '6'; the digit 6 uses a

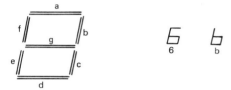

Figure 6.9. The shape of a seven-segment display, with segment letters

horizontal top bar. Most seven-segment displays will also include an eighth segment, the decimal point. This may be placed before or after the segments (*Figure 6.10*) and some care is needed to select the correct type if a decimal point is to be used.

Signals from a BCD counter cannot be applied directly to a seven-segment display, so a BCD-to-seven-segment decoder must be used as an interface. This consists simply of gates

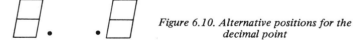

Figure 6.10. Alternative positions for the decimal point

whose inputs are the BCD digits and whose outputs are the driving lines for the seven-segment display. The decoder may also include high voltage outputs or, for some types of display, a.c. outputs. A truth table for a BCD-to-seven-segment decoder is shown in *Figure 6.11*. Decoders which are intended for decimal displays only are gated internally so that the inputs

Count	Segments illuminated
0	a b c d e f
1	b c
2	a b g e d
3	a b g c d
4	f g b c
5	a f g c d
6	a f e d c g
7	a b c
8	a b c d e f g
9	a b g f c

Figure 6.11. Truth table for BCD-to-seven-segment decoder outputs. Depending on the type of display which is to be driven, an illuminated segment may need a 0 or a 1

corresponding to the quantities 10 to 15 give either no output or 'nonsense' outputs. The hex characters cannot be obtained.

Most BCD-to-seven-segment decoders offer other facilities such as lamp test, strobing (see later), and blanking of leading 0s. The lamp-test pin enables all the segments to be lit, so testing that no segments are faulty. Strobing enables the outputs of the decoder to be isolated, so disabling the display. This technique is useful when a display is multiplexed, as described later in this chapter. The leading 0 suppression, or *ripple-blanking*, facility is used when a set of several displays is used. When all the ripple-blanking inputs and outputs are connected in line (*Figure 6.12*), then no 0 will be illuminated ahead of the most significant digit of the display. In this way, the 0s, known as *leading 0s*, in such numbers as 0034 are suppressed, but not the 0s which are part of the numbers, as in 1200.

Decoders also exist for other counting codes such as Gray

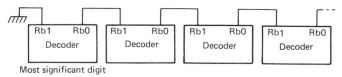

Most significant digit

Figure 6.12. Using the ripple-blanking connections on decoder i.c.s

code and excess-3, but the range of such decoders is rather restricted as compared with binary and BCD decoders.

As mentioned earlier in this chapter, incandescent filament displays can be obtained in seven-segment form, and those have several advantages. They can, for example, be driven from decoders such as the 7448. The most commonly used seven-segment displays, however, are the l.e.d. types. The l.e.d. is basically a diode which emits light from its junction when current flows in the forward direction. The material used is gallium arsenide or indium phosphide, rather than silicon, and the forward voltage is rather high, 2.0 V or more. The maximum reverse voltage permitted is rather low, 3.0 V, and the displays must be arranged so that they cannot be damaged by reverse voltage. Because both TTL and CMOS circuits work with higher voltage levels, a current-limiting resistor must always be wired between each output of the decoder and the corresponding input of the display. The size of this resistor depends on the amount of current which is to flow in each segment, and this amount of current must be less than the maximum allowed by the manufacturer of the display. For example, if a display is to pass 10 mA per segment, and is to operate from a 5.0 V TTL circuit, then the voltage for logic 1 is +5.0 V, and the forward voltage drop across the display is 2.0 V. The 10 mA is therefore driven by the difference of 3.0 V, so that the series resistance, by Ohm's law, is 3/0.01 = 300 ohms. A 330 ohm resistor would be used in practice. A protective diode can be fitted across each display segment if there is any risk of reverse voltage (*Figure 6.13*).

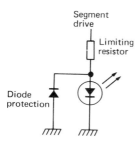

Figure 6.13. Protecting an l.e.d. segment from reverse voltage. This is necessary only when interface circuits with negative supply voltages are used

As well as variation of decimal point position, l.e.d. displays also offer the choice of common-anode or common-cathode connection. A display with a common-anode connection will have one common terminal which is connected to supply +, and the seven cathode terminals (plus decimal point) which are each wired through a limiting resistor to the decoder. The decoder must be a type suited to common-anode displays with a zero output for an illuminated segment, and a 1 for non-illuminated. The common-cathode type of display operates with the common terminal earthed and the segment terminals connected through the limiting resistors to the decoder terminals. In this case, the decoder must be of a type suited to common-cathode displays, giving logic 1 on illuminated segments and logic 0 for non-illuminated segments.

If only the 'wrong' decoder type is available, logic inverters can be used connected between the decoder and the limiting resistors.

Figure 6.14. The 5 × 7 dot-matrix pattern. Figures, letters and symbols can be displayed

The dot-matrix type of display (*Figure 6.14*) permits a complete range of numbers, letters and symbols to be displayed but at the expense of a much more complex decoding system using a read-only memory. As the name suggests, the dot-matrix display consists of dot-sized l.e.d.s arranged in rows and columns, 5 × 7 being a common configuration. The 5 × 7 matrix provides 35 separate l.e.d.s to control, and several types of dot-matrix display have decoding built in rather than provided externally, so reducing the number of leadout wires which would otherwise be needed. Simpler decoding systems can be devised if the dots can be illuminated in sequence rather than simultaneously.

L.E.D. displays of any type have one serious disadvantage, that of high current requirements, which makes them unsuitable for battery-operated displays for viewing over long periods. Calculators making use of l.e.d. displays generally use circuitry which automatically switches off the display (without loss of power to the calculating circuits) after a minute or so of viewing; the alternative to this solution is the use of expensive and sometimes short-lived rechargeable cells.

Low-voltage fluorescent displays are one method of reducing the current taken by a display of several digits. These displays, generally green in colour, are miniature cathode-ray indicators, with electrons from a hot filament being accelerated to a fluorescent anode whose shape is that of a bar of the familiar seven-segment pattern. A voltage of around 18.0 V between anode and cathode is needed for reasonable legibility, and this can be achieved in the way shown in *Figure 6.15*. Since there

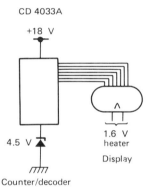

CD 4033A

+18 V

4.5 V

1.6 V
heater

Display

Counter/decoder

Figure 6.15. Driving a fluorescent display from a CMOS decoder

is no noticeable glow at a voltage difference of 4.5 V the cathode voltage of the display can be returned to 4.5 V, using the logic to provide the voltages of 0 V (off) and +13.5 V (on). This enables the fluorescent display to be driven directly from a CMOS decoder counter, such as the CD4026A or CD4033A. If the logic used with the fluorescent display is TTL, then decoders having open-collector transistor outputs (*Figure 6.16*) must be used. The output transistor in the i.c. is formed in such a way as to be able to withstand up to +30.0 V, so that high-voltage displays can be driven directly provided that a suitable power supply is available. The display is the load for the collector connections of the output transistors.

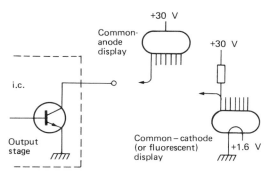

Figure 6.16. Driving high-voltage displays from TTL,
using open-collector outputs

A type of display which nowadays is used very extensively
for battery-operated equipment is the *liquid-crystal display*
(l.c.d.). A liquid crystal is a material whose behaviour is in
some respects midway between that of a liquid and that of a
solid crystal. The important feature of such materials from the
point of view of display techniques is that the material is
sensitive to an electric field applied to it between two conducting plates like the dielectric of a capacitor. The normal
arrangement for an l.c.d. is of a metal or other conducting
material used as a backplate, usually blackened, and a transparent material with a transparent conducting coating, used as
a frontplate. The frontplate can be coated with conductor in
the familiar seven-segment pattern, and then the two plates
assembled into a cell with liquid crystal material filling the
space between the plates. In the unenergised condition the
liquid crystal material is opaque, so the front plate appears to
be reflective. When a voltage, preferably a square wave at a
frequency lying between 30 and 200 Hz, is applied between
the backplate and a selected bar of the frontplate, the liquid
crystal material between these points becomes transparent.
The blackened back surface now appears in the shape of the
segment, in contrast to the reflection from the rest of the display. Since no light is produced by the display, visibility is
good only when illumination is provided. One variation of

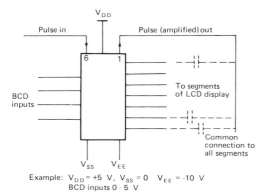

Example: V_{DD} = +5 V, V_{SS} = 0, V_{EE} = -10 V
BCD inputs 0 - 5 V

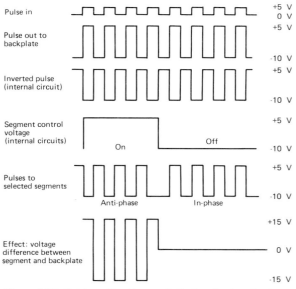

Figure 6.17. Driving a liquid-crystal display (l.c.d.). The driver i.c. provides the amplification and inversion of the pulses and ensures that no d.c. is delivered to the display

the basic scheme, which is not suitable for battery operation, is to make the backplate transparent also, and to have a continuous backlight. When the liquid crystal material has the square wave signal applied, the light from the back of the display shines through to the front in the pattern selected by the decoder which drives the display.

Because the square wave is applied to what is virtually a small-value capacitor, very low currents are required, of the order of 30 μA, for all seven segments of a display. D.C. must not be applied to an l.c.d. so the type of driver circuit needed differs considerably from that used for other displays. The usual scheme is to use a conventional BCD-to-seven-segment decoder action along with a driver stage fed with a square wave input. The square wave is amplified, usually to 15.0 V peak-to-peak, and is fed to the backplate of the display. When a segment is not selected, the signal on the segment is an identical square wave in phase with that on the backplate, so that the voltage difference between the two is zero. When a segment is selected by the decoder circuitry, a square wave which is the inverse of that at the backplate, is applied to the segment, so that the resultant wave between the plates (*Figure 6.17*) is a square wave whose amplitude is equal to that of the supply voltage in each direction. In this way, a 15.0 V supply can supply the equivalent of a 30.0 V peak-to-peak square wave, though the driving square wave input to the i.c. is only of 5.0 V peak-to-peak. A typical decoder driver for l.c.d.s is the CD4055A.

Multiplexed displays

When a display consists of only a few decimal figures, the simplest circuit methods are to use a BCD counter and decoder/driver for each display figure. Displays of four figures or more, however, are usually operated in multiplexed form, in fact the pin connections generally allow only for this form of operation when a complete set of digits forms a single display unit.

A multiplexed display, however many digits it uses, has only one set of segment pins, consisting of the seven bars plus decimal point, eight in all. There must also be a set of common-anode (or common-cathode, whichever is used) pins, one for each display digit. An output on each selected segment line will cause only one digit to be displayed, because only one anode (or cathode) is selected at a time. The multiplexing circuitry must be arranged so that when a digit anode (or cathode) is selected, the segment driving signals are correct for that digit.

One way of using multiplexing on separate displays permits no particular saving on i.c. chips (unless an LSI circuit is used) but a considerable saving in display current, since only one device at a time is switched on. The outline is shown in *Figure*

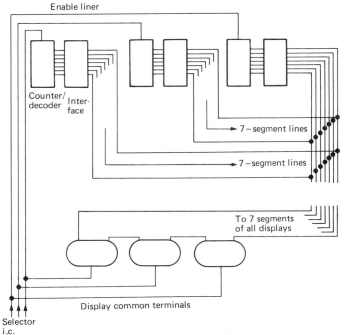

Figure 6.18. A multiplexed display in which the display units are switched in sequence using a common set of seven-segment lines. This is essential if the display is constructed in one block

6.18. Each digit needs a counter/decoder which drives the displays through an interface circuit. Each counter/decoder i.c. has a strobe or enable input which isolates the output until a positive pulse appears at this terminal. To display the least significant digit then, the strobe pulse is applied to this counter unit, whereupon the counter/decoder which has been strobed will release its signals on the output lines, displaying the digit. The switching of the strobe pulses can be carried out by a sequential counter, such as a decoded Johnson counter, or one of the many i.c.s which are available for this purpose. If the decoder has no strobe input, multiplexing can be carried out by applying strobe pulses to a gate capable of driving the common terminal of the decoder.

A more ambitious multiplexing scheme, suitable for the multiplexed type of multi-digit displays, is illustrated in *Figure 6.19*, using a multiplexing i.c. such as the 74153. For example, a set of four one-bit multiplexers can switch a four-bit signal (such as a single BCD digit set) into four output lines from several sets of input lines. Using this scheme only one decoder is needed, fed from the four output lines of the multiplexer; a very useful point if an expensive hex decoder is used.

A display sequence is as follows. For the least significant decimal digit, the control counter pulses the correct display and also selects the correct input set of lines for the multiplexer, so ensuring that the BCD number sent to the decoder is the correct one for the display. The next display is then strobed and the next set of BCD signals sent out on the decoder lines, and so on in sequence.

When multiplexed displays are used, some care must be taken over the frequency at which the displays are strobed. In an l.c.d., for example, there will be a driving frequency of square waves for the display itself, a clock frequency for the circuit logic, and also a strobing frequency for the multiplexer. These frequencies should be related and synchronised, as can be ensured by driving them all from one common source frequency, using divider stages to obtain the various frequencies which are needed. This method avoids the 'beating' effects which might otherwise be noticeable.

124

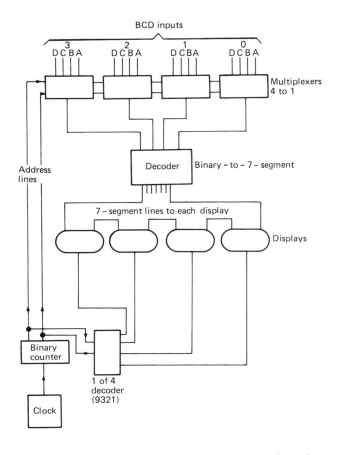

Figure 6.19. A different type of multiplexed display. The binary counter activates address lines which select which digit each multiplexer passes to the single decoder. At the same time the 1 of 4 demultiplexer selects which display is to be used

Video screen displays

The display methods described so far are suitable for calculators and for small hexadecimal displays but not for displaying words arranged in sentences nor for large blocks of figures. For displaying the output of computers a video screen is preferable to any other type of temporary display. Permanent 'displays' such as teleprinters are not considered under this heading.

A video screen display makes use of a TV cathode-ray tube. Although standard methods of scanning need not be used, the availability of low-priced TV scanning components ensures that most video displays use the line-and-field type of scan which is employed by TV receivers. When high resolution is not important, for example when fewer than about 40 characters per line are shown, then a conventional TV receiver can be used as a video display, provided that the signals from the computer (in suitable form) can be modulated on to a carrier in the UHF range of frequencies. The degradation of resolution caused by most modulators and receiver IF passbands makes this method unacceptable for high-resolution units (60 characters per line or more), and a monitor which accepts composite video signals must be used. A mains-operated TV receiver must never be used as a monitor in this way, though some battery-operated receivers can be adapted provided that they are operated only from the low-voltage supplies of the computer.

The next problem concerns obtaining the video waveforms to modulate or feed to the monitor. Unlike a straightforward seven-segment or dot-matrix display, the video screen offers a two-dimensional choice of position for any letter, symbol or number. Because of the wide choice of symbols which can be displayed, simple binary codes are insufficient, and a multi-bit code known as ASCII (American Standard Code for Information Interchange) is generally used. Use of the ASCII code is not universal, but it is by far the most common coding system used on the modern small computers. The ASCII code, illustrated in *Figure 6.20*, encompasses a wide range of instructions as well as symbols. These instructions are designed to cope with the problems of locating characters correctly on the screen, and

Character	Hex	Character	Hex	Character	Hex	Character	Hex
NULL	00						
SOH	01	!	21	A	41	a	61
STX	02	"	22	B	42	b	62
ETX	03	⇌	23	C	43	c	63
EOT	04	$	24	D	44	d	64
ENQ	05	%	25	E	45	e	65
ACK	06	&	26	F	46	f	66
BEL	07	'	27	G	47	g	67
BS	08	(	28	H	48	h	68
HT	09	)	29	I	49	i	69
LF	0A	*	2A	J	4A	j	6A
VT	0B	+	2B	K	4B	k	6B
FF	0C	,	2C	L	4C	l	6C
CR	0D	−	2D	M	4D	m	6D
SO	0E	·	2E	N	4E	n	6E
SI	0F	/	2F	O	4F	o	6F
DLE	10	0	30	P	50	p	70
DC1	11	1	31	Q	51	q	71
DC2	12	2	32	R	52	r	72
DC3	13	3	33	S	53	s	73
DC4	14	4	34	T	54	t	74
NAK	15	5	35	U	55	u	75
SYN	16	6	36	V	56	v	76
ETB	17	7	37	W	57	w	77
CAN	18	8	38	X	58	x	78
EM	19	9	39	Y	59	y	79
SUB	1A	:	3A	Z	5A	z	7A
ESC	1B	;	3B	[	5B	{	7B
FS	1C	<	3C	\	5C	|	7C
GS	1D	=	3D	]	5D	}	7D
RS	1E	>	3E	↑	5E	~	7E
US	1F	?	3F	←	5F	Delete	7F
SP	20	@	40	\.	60		

they take the form of carriage return, line feed, and similar instructions which would also be used for a teleprinter. The ASCII code also allows for a parity bit, an extra bit added to the seven-bit code to make the total number of 1s even. By using this additional parity bit, an error in a coded character can be detected and signalled by the computer system simply by counting the number of 1s in each coded character.

Conversion of the ASCII code to video signals requires a considerable amount of digital circuitry which may need a large number of chips or the use of a microprocessor (next chapter). Multi-chip logic systems permit some control over the number of characters which are displayed per line, whereas single chip logic systems usually fix this quantity at an amount which is suitable for professional-quality video monitors. The term 'single chip' is rather misleading in the sense that several support chips are usually needed in addition for a complete video screen logic decoder, and these raise the price of the decoder to more than that of the 'multichip' version. As might be expected, the decoding procedure is very complex and is beyond the scope of this book.

Instructions

			SI	– shift in
SOH	– start heading/message		DLE	– data link escape
STX	– start text		DC 1-4	– device control signals
ETX	– end text		NAK	– negative acknowledge
EOT	– end transmission		SYN	– synchronous idle
ENQ	– enquiry		ETB	– end transmission block
ACK	– acknowledge		CAN	– cancel
BEL	– ring bell		EM	– end medium
BS	– backspace		SUB	– substitute
HT	– horizontal tabulation		ESC	– escape, prefix
LF	– line feed (new line)		FS	– file separator
VT	– vertical tabulation		GS	– group separator
FF	– form feed (next page)		RS	– record separator
CR	– carriage return		US	– unit separator
SO	– shift out		SP	– space

Figure 6.20. The ASCII code. To save space the hex equivalent of the seven-bit binary code has been shown. Not all of the instruction codes may be needed by a given teletype

7 Microprocessor systems

Any truth table whose action can be carried out by the usual combination of logic gates and shift registers can also be carried out by a computer program. Such a program operates by examining the inputs *in sequence* and producing an output which depends on what inputs have appeared. Whereas the use of gates, registers and similar digital circuits connected together is called a *hardware solution*, the use of a program is called a *software solution* to the problem of carrying out (*implementing*) a truth table.

The first true microprocessor, the INTEL 4004, appeared in 1969, and it was an attempt to produce the logic system of a computer, with the capability of being programmed, on a single i.c. chip. Such a chip must contain sufficient gates to produce the basic gate functions such as AND, OR, X-OR, and the arithmetical functions such as add, complement-and-add, plus shift registers to retain the binary numbers, in eight-bit bytes which are being processed, and counters to control the sequence of events during a program. Since these are the functions carried out by the *Central Processing Unit* of a large computer, the microprocessor chip is often referred to as the CPU.

At the time of writing, most semiconductor manufacturers make CPU chips either of their own design or identical 'second-source' copies of other CPUs. Though these CPUs vary considerably in the way that they carry out instructions, a number of features common to all the present generation of CPUs can be summarised.

To start with, these microprocessor CPUs all use eight-bit

bytes, meaning that eight bits of digital signals can be processed in parallel. At the time of writing, 16-bit microprocessors are beginning to appear, but most of the present applications can be carried out comfortably with eight bits — the difference is of processing time only.

Any CPU chip must therefore have provision for the input and output of an eight-bit byte into or out from an eight-bit shift register. The main register used for the reception or transmission of these eight-bit bytes is called the *accumulator* and all the results of processing operations are also held in this register. A few designs can use this register only for loading or dumping data bytes, other CPUs can be instructed to use other registers.

Loading a byte into a shift register requires eight input pins (for parallel loading) and unloading might be expected to use another eight pins, 16 in all. Because of the need to keep the pin-count down, however, and to ensure a simple circuit lay-out, it is convenient to use only eight pins for both inputs and outputs. A set of internal gates controlled by an internal signal connect the eight pins of the chip either to the register input or output as required. The signal which controls this dual use of the data pins must be arranged so that the input gates are selected unless program instructions specify an output. A few designs use a pin input for a signal which isolates the CPU from the data pin voltages. All designs, however, have provision for the output of a signal which can be used to control other chips by indicating whether data is being input to the CPU or output from the CPU. In all cases the input of data to the CPU is referred to as 'reading', and the output as 'writing'.

Before we can proceed with a description of the CPU, we need to look more closely at the chips surrounding it. Without these chips, the CPU cannot function in any way, so that we cannot describe the operation of the CPU without also describing the action of the chips on which it depends.

Of these other chips, memory devices are the most important. A memory chip stores digital bits until they are needed by the CPU, and the act of supplying bits (reading from memory) does not destroy or modify the stored data digits. A memory may be either volatile or non-volatile. A volatile memory is one which loses its stored information when the power supply is switched off. A shift register is a simple form of volatile memory, since the flip-flops will not remain switched in the absence of a power supply. In addition, when power is switched on again the flip-flops will switch over randomly so that meaningless digits are stored. Non-volatile memories, on the other hand, can be used for permanent storage because the data bytes stored in these memories are neither erased nor modified when the power supply is switched off.

As in a large computer, the non-volatile memory in a microprocessor system may be a tape, cassette or magnetic (floppy) disc store, but the microprocessor, unlike large computers, also makes use of non-volatile memory chips known as ROM, a name formed from the initial letters of *Read-Only Memory*, because these memory chips are so designed that the stored bytes cannot be altered by any action of the micro-processor. Unlike the magnetic tape or disc form of non-volatile memory the data stored in a ROM can be obtained and read out to the CPU very quickly, usually in a time of less than a microsecond. The main use of ROMs in a microprocessor system is to provide a program, called a *monitor* or *bug*, which enables the system to be used and further programmed. Without such a monitor a very much greater quantity of logic chips are needed to allow the microprocessor to carry out the input and output of data.

A ROM such as a monitor program or a character generator (for storing the instructions needed to print characters on to a TV screen), which is a widely used and essential part of a microprocessor assembly, will be mask-programmed. This implies that the ROM is manufactured with the program in place in the form of connections made inside the i.c. This process is prohibitively expensive for small quantities since the cost of the masks (the jigs for i.c. manufacture) has to be paid. When these fixed costs can be spread over a large number of chips, however, mask-programmed ROMs become very much cheaper than other types.

For any application of a microprocessor, a program will have to be worked out and tested and this can be done by using volatile memory in the first instance, but if the microprocessor is to be used in a control system the program will have to be produced as a ROM. There are several ways of doing this. The program may be written for a full-sized computer and then tested on the micro system using volatile memory. The main computer program can then be printed out and sent to the semiconductor manufacturer who will produce ROM masks directly from the program. Another method, which is necessary if no mainframe computer is available, is to use a type of ROM known as a PROM, which can have a program written into it. The letter 'P' in PROM signifies *Programmable*, but the programming cannot be carried out by the normal action of the microprocessor circuits. The advantage of using a PROM is that prototypes of a microprocessor control circuit can be made up and thoroughly tested before mask-programmed ROMs are used. Also, for small production runs the use of PROMs is usually more economical than the use of mask-programmed ROMs.

Fusible-link PROMs contain a set of connections which, like fuses, can be broken by passing a current of about 200 mA at 17.0 V. This is pulsed, and the voltages on other parts of the PROM circuits are arranged so that no damage can occur elsewhere. As manufactured, each link of the fusible-link PROM is unbroken, so that each memory bit is a logic 1. The act of programming is therefore to fuse each link where a logic 0 is needed. Once fused, of course, the links cannot be reset, though more links can be broken if need be. The current needed to fuse the links is very much greater than the microprocessor could possibly supply, if by chance the PROM were connected to a data pin set to logic 0, so separate programming circuits are needed. These circuits are arranged so that the program can be read from a volatile memory or from a tape store into the PROM, and sufficient current passed to fuse a link for each 0.

A later step in the development of PROMs has been the Erasable PROM, or EPROM. The most popular system at present uses ultra-violet light to erase the PROM, which has a quartz 'window' covering the chip. Erasure results in each part of the memory storing logic 1 and is carried out by leaving the window of the chip exposed to a UV tube for 30 minutes or so at a distance of a few centimetres. The programming operation sets selected bits to logic 0, and this is done by holding selected bits, in sequence, at logic 0 for one millisecond and cycling around all the bits 125 to 150 times to fix the program. During programming, much higher supply voltages are in use, so that normal operation has no effect on the PROM thereafter.

The volatile memory i.c.s used along with microprocessors are known as *Random Access Memory,* or RAM, an old name once used to distinguish a memory of this type from one in which bits were fed out in sequence. Either type of memory can be made by using flip-flops arranged as registers, but since all the memory chips now used with microprocessors permit random access (meaning that a bit can be selected at random from any part of the memory), whether RAM or ROM, the term *Read-Write Memory* would be more appropriate.

Unlike ROM, the RAM can be written into from the microprocessor by using a 'write' control voltage. One logic signal level on the write pin of the RAM i.c. will permit the microprocessor to store data bytes into the memory, and the opposite signal will act to switch the memory to 'read' so that data bytes are read out from the memory to the microprocessor. The write process automatically results in replacing any previously-stored information, but the read process is non-destructive.

Both static and dynamic RAMs exist. The dynamic type relies on charging MOS capacitors which, because of the loss of charge caused by leakage, have to be recharged at intervals. This is done by a 'refresh' cycle in which a refresh pulse is routed to each capacitor storing a logic 1. Any system making use of dynamic memories (which are relatively inexpensive and consume very little current) must therefore provide for a refresh pulse to maintain storage. Obviously this refresh pulse must occur at a time when the memories are not being written nor read, so it has to be under the control of the microprocessor CPU. The Z80 CPU carries this to the logical conclusion by having the microprocessor itself generate the refresh pulses for any dynamic memory.

Static memories, using flip-flops or similar circuits for storage, need no refresh pulses and are therefore useable in microprocessor systems where the additional refresh pulse circuits would be unwelcome. At the time of writing, the cost (and power consumption) of a static memory RAM is considerably higher than that of an equivalent dynamic RAM.

Whatever type of memory is used in the microprocessor support circuits some provision has to be made for selecting the data to be read out from the memory, or for selecting a part of the memory into which data can be written. The selection of a data bit or set of bits is carried out by using a set of address lines on which digits are used to specify a code number or address for each byte in the memory. A decoder within the memory chip is activated by the signals on the address lines so as to select one part of the memory to connect to the output pins.

The nature of this decoding depends to a large extent on the way in which the memory chip is organised. For example, a chip described as a 1024×1 bit memory will be able to store a single bit at each of 1024 different addresses, so that each bit can be obtained from a single output/input pin. To address a total of 1024 (2^{10}) bit positions ten address lines must be used. For a microprocessor which uses an eight-bit byte, eight such memory chips would be used (*Figure 7.1*), each with its readout

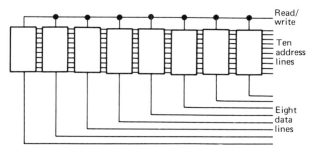

Figure 7.1. Using one-bit memory i.c.s. One memory i.c. is needed for each of the eight data bits. For the example shown, which uses 1024-bit chips, 10 address lines are needed and these are connected to each chip

pin connected to one of the data pins of the CPU. The ten address pins of the memories would be connected in parallel to be driven by the address lines of the microprocessor. Only 10 of the usual 16 address lines of the microprocessor would need to be used for this example.

Other memory organisations, such as 512×4 or 256×8, use fewer address lines but more data lines, the particular method of organisation used depending on what hardware is available, comparative prices, and the need for a given quantity of memory. Some gating circuits may have to be arranged so that each memory is correctly addressed.

The addressing of memory is carried out by the micro-processor CPU under the direction of a program or by counting up binary numbers on the address line. The standard number of address lines for the CPU is 16, as already mentioned, which permits the use of 2^{16} (65 536) memory addresses. Since no memory chip freely available at present requires 16 address lines some selection method is needed, and this is provided by a chip-select pin on each memory chip. Imagine, for example, a 128×8 memory chip, such as the popular 6810. This can be arranged to read out to the microprocessor on its eight data lines any one of the 128 bytes which can be addressed by seven address lines when one of the chip select pins is at logic 1. With the chip-select pin at logic 0, the memory chip is isolated, floating, neither address nor data signals affecting it. The chip-select pin can be controlled by a gate so that if the number on the address lines is greater than 128, a 1 on any of the lines from the eighth to the 16th indicates that this chip is not selected. Another chip of the same type can be connected now to the same set of seven address lines, but with its chip-select pin controlled by another gate so that the pin voltage is 1 when the eighth line of the address lines has a 1 present. A third similar chip can be selected when the ninth address line is at 1, and so on.

A few microprocessors do not use the full 16 address lines as pin outputs from the CPU. One notable design, the National Semiconductor INS8060, uses 12 pins for the lower 12 bits of address, with four more address bits available by gating the data lines with a pulse which is available at another pin. Another type, the Fairchild F8, has no address pins on the CPU but uses a separate memory chip with addressing and counting built in.

The full complement of 16 address pins and eight data pins, numbered starting from 0, requires connections between the microprocessor CPU and all the surrounding chips. The connections are made on a *bus*, a set of parallel lines on a printed circuit board, or a set of cable wires linking corresponding pins on each chip. The data bus consists of the eight lines, numbered D0 to D7, which transfer data bits to and from the microprocessor; the address bus consists of the 16 lines, numbers A0 to A15, which output address information to the memories and other chips. These connections account for 24 of the normal complement of 40 CPU pins.

Of the remaining CPU pins, some are used for power supplies and clock signals. Early CPU designs used several power supply lines at different voltages; for example, the 8080 used +12.0 V, +5.0 V and −5.0 V with a common earth. Later designs have tended to require only one voltage supply, usually +5.0 V, so that only two power supply terminals, + and earth, are needed, and this has the advantage of releasing more of the 40 pins for signals. Since all microprocessor operations are synchronous, a clock pulse is needed. A few microprocessors have internal oscillator circuits so that clock pulses can be generated by connecting a frequency-sensitive circuit network between two pins. The network may be as simple as a resistor and a capacitor for clock pulses which do not have to be held to a constant frequency, or may be a quartz crystal operating at 1 MHz or more. The use of an external clock pulse generator is more common, and requires only one input pin if a single-phase clock is used. A few designs need a two-phase clock, and one chip needs a four-phase clock.

The remaining pins of the CPU are used for control signals which are also connected into a bus wiring system. The signals along this bus vary greatly from one CPU type to another but some factors are necessarily common to all types. Signals controlling the reading from or writing to memory chips must be part of the control bus, and these are referred to as *data-*

bus definition signals since they decide which way signals are flowing along the data bus lines. Timing signals allow the CPU to wait until data is ready to be delivered. A 'ready' signal input, for example, applied to the CPU indicates that a chip has data ready to put on to the bus. This is necessary because there is always a time delay between the output of an address from the address pins of the CPU and the output of data from other chips, and this time delay may vary from one type of chip to another. For a short time after an address has appeared on the address lines, therefore, the signals on the data bus will fluctuate as the memory gate circuits operate. Until these fluctuations have ceased and the signals are steady no information should be read in, and the ready signal is used to indicate this state.

A 'hold' signal has the effect of disconnecting the CPU from the data lines and address lines so that other circuits can control these lines. During the hold period the CPU waits and outputs a signal on another pin as an acknowledgement that normal service is suspended. The ready and hold signals are used on the 8080 CPU, but other CPU chips use timing signals which are differently named and which in some cases have rather different actions. All CPUs have, however, a reset pin which clears the program counter register (see later) and so enables a program to start at some pre-determined memory address.

The third set of control signals is concerned with *interrupts*. Microprocessors operate synchronously with a fixed clock cycle but may have to read in signals from other devices, such as analogue-to-digital converters, which are not clocked at the same frequency, or from devices, such as logic gates, which are not clocked at all. The method used to allow for such inputs, which may occur at any point during a program, is called an *interrupt*. The method of carrying out an interrupt varies from one CPU design to another but basically the scheme is that the device with a signal byte to deliver sends out a logic signal called an *interrupt request* which may, for example, be a logic 0 on a line which is normally set at logic 1 and which is connected to an interrupt request pin on the CPU. At the end of the program instruction which the CPU is carrying out at the time, the voltage on another CPU pin, the *interrupt acknowledge*, is changed. This is a signal that the device which caused the interruption can go ahead and deliver its data. At the same time, the CPU must now switch to a program which will deal with the interrupting signals – this is called an *interrupt service routine*.

If no interrupt service routine has been prepared by the programmer then the microprocessor will stop operating, since it has no instructions to deal with the interrupt. The start of the interrupt service routine must include instructions to preserve the contents of the CPU registers which existed at the time of the interrupt, so that the normal program can resume. The end of the interrupt service routine must include program instructions which will replace all these register contents for the normal program and then set the normal program going again.

Most microprocessors allows interrupts to occur only after specific parts of the main program, using an instruction called *interrupt enable*, and can prevent an interrupt from itself being interrupted by using a program instruction called *interrupt disable*.

Inputs and outputs

So far, we have dealt mainly with signals passed between the CPU and its memories, but since the purpose of the microprocessor is to control machinery by means of sending out digital signals, some output signals must exist. Similarly, some provisions must be made for the input of programs other than the monitor program which is in ROM, data, and instructions such as 'go', 'halt', 'reset'. These signals are dealt with by chips described as PIAs or PIOs – Peripheral Interface Adaptor or Peripheral Input/Output.

Techniques vary considerably from one microprocessor to another but one very common system is to have connected to the data bus lines a PIA chip which is switched either by control signals or by a combination of address and control signals.

When the PIA is selected, either by an address as if it were part of the memory or by chip-select signals, then it makes gate connections to the data lines. In addition, another control signal will determine whether the PIA is to act as an input or an output terminal, or *port*. If the PIA is selected to act as an output then the signals present on the data lines at that time will also appear on the eight output pins of the PIA. One PIA chip usually has two sets of data input/output pins, and as these can be used as inputs or outputs in any combination a group of eight pins could control or accept signals from eight devices needing only two-level (0 or 1) signals. The selection of the PIA is entirely under the control of the CPU program, so that such elementary steps as using a keyboard input or writing out data to a set of displays are impossible unless a monitor program exists to instruct the CPU accordingly. A few types of CPUs have built-in PIA facilities, but this is unusual.

CPU action

As a full description of the action of any type of CPU would have filled the whole of this book, only a brief general description will be attempted here. The microprocessor CPU is not a simple device, and simple explanations are usually rather vague and misleading unless the reader has a fairly extensive experience in digital electronics.

The CPU has within the chip several registers of the parallel input, parallel output type, some of eight bits and some of 16 bits. Two particularly important registers are the accumulator, to which the results of all calculations are returned, and the program counter, which controls the progress of the CPU program from one instruction to the next. All registers are, of course, controlled by the clock pulses, but this does not mean that each program step is carried out in the time of one clock pulse. It is much more usual for a large number of clock pulses to be needed to carry out one program step, as many as 88 clock pulses for some operations, because several actions take place in sequence within the chip for each program step, and each separate action of a step will need one or more clock pulses.

Imagine a microprocessor set up with ROM, RAM and PIO (*Figure 7.2*). The contents of the ROM will be the monitor program, and the address of the start of this program will be binary 0000000000000001 (the leading 0s will be omitted in future references), so that the CPU will always start from this

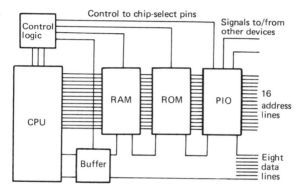

Figure 7.2. An elementary microprocessor system. The ROM contains the program, and the RAM is used for storing input information and for temporary items (scratchpad use). The PIO connects with external devices such as keypads or displays. A buffer i.c. is shown in the data lines to isolate the CPU from the memories and PIA when necessary, though this is not always needed

point after having been reset. When power is switched on, the circuits around the CPU operate the reset pin, and the reset action continues for several clock cycles, giving the clock time to settle down to steady oscillation and also allowing all the registers to clear. A time of 500 clock pulses is often allowed for this resetting operation. This will not necessarily clear the RAM, so that each address in RAM will contain a byte of random digits (*garbage*).

When the reset is released, the program counter register moves from 0 to 1 (leading 0s omitted here), and so selects the starting address of the ROM chip which provides the operating program. Each instruction of the operating program will require several clock pulses to complete, and may involve the transfer of data between the accumulator register and memory positions or the in/out ports. After each instruction has been completed the program counter is incremented (increased by 1).

The monitor program is usually arranged to cycle continuously in the absence of other instructions so that a keyboard connected to an input port can be used to generate a RAM address and then place a data byte at that address. As the clock rate of the microprocessor is so fast that many cycles of program could be completed in the time for which a key on the keyboard is pressed, keyboards are usually connected so as to operate by an interrupt system. Rather than use decoding on the keyboard (converting hexadecimal keys into binary code) the decoding is usually carried out by the microprocessor under the control of the monitor program. Similarly, the display of hexadecimal characters on seven-segment displays is carried out by programming rather than by the use of decoder chips. A good monitor program enables the user to inspect each step of a program written into RAM, to check the contents of each RAM memory position, to reset to the starting point of a RAM program, and to modify any program as required. In addition, when the starting point of the program in RAM is selected, a go key should enable the microprocessor to run the program, and a reset key should make it possible to return to the starting conditions if the RAM program is faulty. A very useful facility is that of going through a program step-by-step instead of at the full normal clock rate.

The scheme just described is of a minimal system containing just about the minimum number of components to be able to write simple programs. Writing a program for a microprocessor is extremely tedious even when hexadecimal characters are used, and these steps will not be described here, but the addition of more peripheral chips can make the task of programming considerably easier. One very useful chip is a serial converter which converts the eight-bit byte form into serial form, one pulse at each time, thus enabling information to be recorded on tape, sent to a teleprinter or to a TV terminal and so displayed (if suitable decoding exists). Using a complete letter/digit keyboard which outputs binary numbers in the standard ASCII form, the task of programming can be carried out in a higher level language.

This imples that the instructions can be typed in the form of words and numbers, following a pattern which is easier to recognise than the binary numbers of machine code or their hexadecimal equivalents. Since the microprocessor still operates in binary code, however, a program must be used to convert each instruction from the keyboard into the correct set of machine-code numbers. The next step up from machine code is an assembler program which enables short mnemonic code words to be keyed in, rather than binary or hexadecimal. Such an assembler program occupies only a comparatively small amount of ROM, because assembly language is reasonably close to machine code — it's a matter of converting an instruction such as LDI (Load Immediate) into the machine code, e.g. 11000100. The next stage up is a compiler or interpreter program; this allows the use of a computing language such as BASIC (Beginners' All-purpose Symbolic Instruction Code), FORTRAN or ALGOL. A compiler program takes up a very large amount of memory space, ROM if the compiler program is wired in or RAM if the compiler program is recorded on tape to be loaded in each time the program is to be used. For applications of microprocessors which require continual keyboard work (such as teletype terminals, word processors), such compilers are necessary, but for the development of simple machine-control systems a relatively simple assembler is more useful.

To complete this necessarily brief description, the software side of microprocessing must be mentioned. Until it has been programmed, a microprocessor system is quite useless. For each CPU there is a software instruction set — a list of the machine codes which must be used as instructions to carry out program steps. When a program has been written these code bytes will be fed into the data bus from the program memory each time the program counter is incremented, and the action which is called for by that step of the program will be carried out before the program counter is incremented again.

With a few exceptions, the instruction sets of CPUs vary very considerably, so that it would be almost impossible to memorise more than one. The notable exceptions are the 8080 and Z80, because the letter is a development of the 8080 and uses all the instruction sets of the 8080 plus many more of its own. An instruction generally consists of one or more bytes which may be followed by one or more bytes of data. The coding of the instruction bytes indicates to the CPU whether any further instruction or data bytes are to follow.

By way of example, the 'halt' instruction does not require any following byte, but an instruction such as LDI, meaning load into the accumulator the byte contained in the next memory position, must be followed by a byte of data, otherwise the instruction cannot be carried out. The byte following the instruction code may, however, complete the instruction in other ways, each of which is planned by the first byte of the instruction code. For example, the second byte may specify a number of program steps forward or backward from the program count, so that the program counter shifts to a new number. Another possibility is that the byte following the instruction byte may be part of an address in memory, and the number stored at that address may provide the next instruction, or be used in a calculation according to the coding of the first instruction byte.

For details of microprocessor programming the reader should consult a programming manual, preferably one of the excellent manuals which are supplied with microprocessor development units such as the KIM-1. Regrettably, not all manuals or even manufacturers' information bookets are of such a high standard, and a lot of 'unofficial' booklets on programming are neither well printed nor well explained for the beginner.

Index

MICROPROCESSORS FOR HOBBYISTS
RAY COLES

Based on two popular series in Practical Electronics,
Microprocessors for Hobbyists is an introduction to
this fascinating and challenging field, which will
appeal to businessmen and computer students, as well
as to electronics hobbyists.

92 pages   0 408 00414 2

INTRODUCTION TO MICROCOMPUTER PROGRAMMING
PETER C. SANDERSON

A practical guide to programming, with self-test
exercises, to enable microcomputer owners to make
the maximum use of their machines. Describes Basic
(including common variants) and assembly languages
of microcomputer systems commonly available.
Designed for those with no programming experience,
including home users, teachers and owners of small
businesses.

144 pages   0 408 00415 0

Newnes Technical Books
Borough Green, Sevenoaks, Kent TN15 8PH

─Beginner's Guides─

. . . provide a basic understanding of a wide range of subjects. If you would like to know more about these and other titles in the series, please write to the address below.

BEGINNER'S GUIDE TO TRANSISTORS — 2nd Edition
Ian R Sinclair and J A Reddihough

Shows how the transistor is used in everyday practical circuits, and introduces the various techniques involved. The approach throughout is non-mathematical and the emphasis is on circuit operation rather than design.

162 pages   0 408 00374 X

BEGINNER'S GUIDE TO INTEGRATED CIRCUITS — 2nd Edition
Ian R Sinclair

This book is for the comparative newcomer to electronics, with some knowledge of transistor circuits, wishing to move onto an understanding of i.c.s. Many examples are given of practical i.c. circuits. Linear, digital and ULA i.c.s are covered, and there is a brief introduction to digital circuit techniques. The operation and uses of several specialised types of i.c. are also described, and the second edition includes a brief description of the microprocessor and allied chips.

192 pages   0 408 01301 X

BEGINNER'S GUIDE TO COMPUTERS — 2nd Edition
T F Fry

An insight into what lies behind the computer end products that affect us, explaining in layman's terms what a computer is and how it sets about doing what is required of it. Reading this book will stimulate an interest in the subject and so be the stepping-off point for a more detailed study of what has become one of the most important and far-reaching areas of technology in our society today.

190 pages   0 408 01307 9

BEGINNER'S GUIDE TO AUDIO
Ian R Sinclair

For the beginner with an interest in sound reproduction who wishes to know more about the methods involved and the electronic circuits used. Systems such as disc recording, tape recording and stereo radio are clearly explained.

192 pages   0 408 00274 3

Newnes Technical Books
Borough Green, Sevenoaks, Kent TN15 8PH

Beginner's Guides

BEGINNER'S GUIDE TO RADIO — 8th Edition
Gordon J King

Introduces all aspects of radio technology, from simple electromagnetic theory, through the full range of radio components and circuits, to the sound you hear from the loudspeaker. The eighth edition is completely rewritten and updated.

240 pages   0 408 00275 1

BEGINNER'S GUIDE TO TELEVISION — 6th Edition
Gordon J King revised by **E Trundle**

An introduction to the technical aspects of television broadcasting and reception, will prove valuable to radio students and technicians, amateur experimenters and constructors, dealers and salesmen and all who are seeking a clear, concise and non-technical explanation of the subject.

236 pages   0 408 01215 3

BEGINNER'S GUIDE TO COLOUR TELEVISION — 2nd Edition
Gordon J King

A guide to the principles of the NTSC and PAL colour television systems, and the method of operation of the PAL system, from aerial to display tube.

202 pages   0 408 00101 1

BEGINNER'S GUIDE TO TAPE RECORDING
Ian R Sinclair

Presents the principles and techniques of tape recording, the choice and use of machines (both reel-to-reel and cassette), transcriptions from radio, disc or tape, mixing, editing, sound and picture synchronisation and nature recording.

176 pages   0 408 00330 8

Newnes Technical Books
Borough Green, Sevenoaks, Kent TN15 8PH

Questions and Answers

Each book contains simple and concise answers to questions which puzzle the beginner and the student – from first principles to a useful level of practical knowledge. Why not write *today* for further information?

QUESTIONS AND ANSWERS ON INTEGRATED CIRCUITS
– 2nd Edition
R G Hibberd

Covers thick and thin film, monolithic and hybrid, digital and linear i.c.s and also deals with Boolean algebra and binary notation. Resistor, diode and transistor logic circuits are described and compared and typical applications are discussed.

112 pages   *0 408 00466 5*

QUESTIONS AND ANSWERS ON AMATEUR RADIO
F C Judd G2BCX

Explains amateur radio in simple terms – what it is, how it started and how it has developed. Covers all aspects of transmission and reception, the radio amateurs' examination and explains simply how *you* can become a radio amateur.

120 pages   *0 408 00439 8*

QUESTIONS AND ANSWERS ON RADIO REPAIR
Les Lawry-Johns

Offers practical advice on dealing with commonly found faults in radios. Possible faults are presented and guidance given, using a minimum of tools, to enable such faults to be located and rectified. The approach is non-technical and assumes some knowledge of basic theory.

96 pages   *0 408 00367 7*

Newnes Technical Books
Borough Green, Sevenoaks, Kent TN15 8PH

Questions and Answers

Each of these books contain simple and concise answers to questions which puzzle the beginner and the student — from first principles to a useful level of practical knowledge. Why not write for further information today?

QUESTIONS AND ANSWERS ON ELECTRONICS — 2nd Edition

Ian Hickman

This book provides a condensed account of a wide-ranging subject, intended to give the interested layman and the student an insight into the underlying principles and numerous applications of electronics.

160 pages   0 408 00578 5

QUESTIONS AND ANSWERS ON COLOUR TELEVISION — 3rd Edition

E Trundle

A simple practical account of colour television transmission and reception for the enthusiast, technician and service engineer. Now thoroughly revised to cover new tube technology, modern methods of high-voltage generation, beam current limiting, flyback blanking, user controls, pincushion correction, integrated circuits, decoding techniques and modern line timebases.

136 pages   0 408 01305 2

QUESTIONS AND ANSWERS ON TRANSISTORS — 4th Edition

Ian R Sinclair

An easily readable introduction to the basic features of transistors and related devices (diodes, thyristors, etc.), how they work and what they can do, with a survey of the many applications in which they are used. There are also hints on servicing transistorised equipment. The text has been updated and extended for this new edition.

112 pages   0 408 00485 1

Newnes Technical Books
Borough Green, Sevenoaks, Kent TN15 8PH